Chemical Processing Nomographs

Chemical Processing Nomographs

edited by

Dale S. Davis

Consulting Editor, Chemical Processing

and

Raymond A. Kulwiec

Associate Editor, Chemical Processing

2nd edition, revised and enlarged

CHEMICAL PUBLISHING COMPANY·NEW YORK ·1969

1st edition 1960
2nd edition 1969

© 1969

Chemical Publishing Co. Inc.
New York, N. Y.

Preface

The initial publication of *Chemical Processing Nomographs* in 1960 filled a long-felt need for a bound collection of the highly popular nomographs that appear regularly in *Chemical Processing*, published by Putman Publishing Company, Chicago, Ill. Designed to help managers achieve speedy and effective solutions to plant operating problems in the chemical-processing industries, these charts have continued to receive enthusiastic reader acceptance.

Since the first edition, requests have come in with increasing frequency for a second, more current collection. This present volume is our happy response to this demand. It contains one hundred and thirty-six of the most helpful nomographs that have appeared in *Chemical Processing* since January 1959 as well as eighteen from the first edition. This collection represents the individual efforts of nomographic contributors throughout the United States and abroad. Without this stable of "nomogramaniacs" this volume would not have been possible.

We trust that our old friends in industry will find this second edition as useful as the first. For those just now discovering *Chemical Processing Nomographs*, we hope the acquaintance will be a lasting one, and that these charts will become helpful, on-the-job allies.

Dale S. Davis

Bailey Island, Maine
January, 1969

Contents

UNIT 1

Flow of Fluids

Flow of Liquids—Flow of Gases—Related Topics

The study of the flow of fluids has important industrial applications, many of which are given here. Related topics deal with ventilation, corrections to rotameters, equivalent diameter, hydraulic radius, and velocities required to fluidize solids. Throughout, the charts are based on definite equations or on well established data.

1-1   Flow Rate For Bleach Liquor

SHIRLEY E. HENRY

Figure 1-1, used to compute the rate of flow of hypochlorite bleach liquor required to bleach paper pulp, is based on the equation

$$V = \frac{39.95 \ RP}{B}$$

where V = required rate of flow of hypochlorite bleach liquor, gal/min
    R = pulp bleaching rate, tons/hr
    P = chlorine required, as percentage of pulp
    B = concentration of hypochlorite bleach liquor, grams available chlorine/liter

Typical Example

    R = 10.6 tons of pulp bleached/hr
    P = 3.4% chlorine
    B = 31.0 grams of available chlorine/liter

Connect 10.6 on the R-scale and 3.4 on the P-scale with a straight line. Note where this line crosses the α-axis. Extend straight line from 31 on the B-scale to intersect the V-scale at 46.4 gal/min.

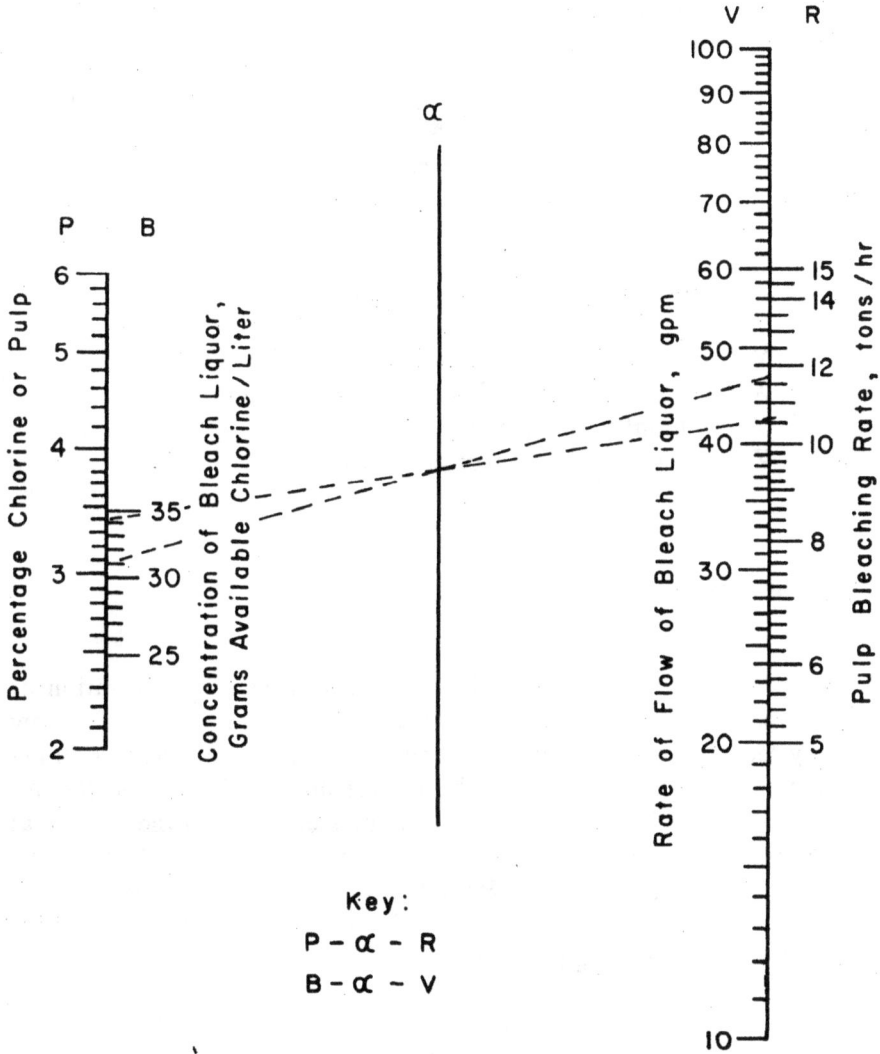

Figure 1-1

1-2   Chemical-Addition Flow Rates

JAMES T. WILLIAMS

Figure 1-2 for chemical-addition flow rates in a paper mill is based on the equations

$$Z = \frac{PW}{6000} \tag{1}$$

$$V = 3.785\frac{Z}{X} \tag{2}$$

$$Y = \frac{60X}{Z} \tag{3}$$

where P = percentage of chemical added
       W = paper production, lb/hr
       V = rate of flow of chemical solution, liters/min
       X = concentration of chemical solution, lb/gal
       Y = number of seconds for addition of 1 gal of solution
       Z = rate of addition of chemical, lb/min

Typical Example

What rate is required to add 0.5% chemical at a concentration of 0.2 lb/gal to a paper machine producing 7800 lb/hr of paper? How many seconds would be needed to supply 1 gal of solution? Following the key, connect 0.5% on the P-scale and 7800 lb/hr on the W-scale with a straight line and note the intersection with the Z-scale at 0.65 lb/min of chemical. Connect this point and 0.2 lb/gal on the X-scale with a straight line. Read the rate of flow of solution as 12.3 liters/min on the V-scale. Note on the Y-scale that 18 sec would be required to supply 1 gal of solution.

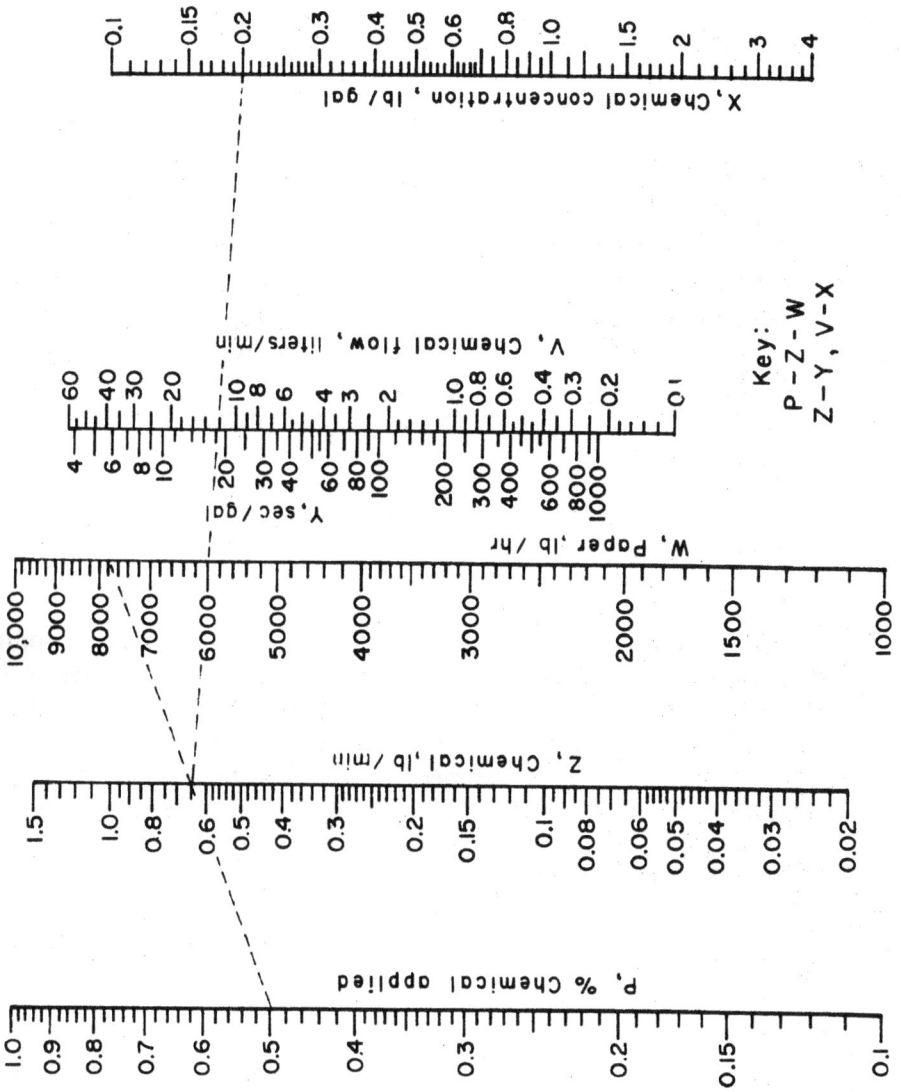

Figure 1-2

1-3   Flow Rates in Mercury Cells

R. W. RALSTON

For mercury cells (caustic-chlorine cells),

$$R = \frac{3.00\ I}{\Delta Na}$$

where R   = rate of flow of mercury amalgam, lb/min
     I   = cell current, kiloamp
   ΔNa = lb sodium pickup/100 lb amalgam

The quantity ΔNa, or % Na pickup, is equal to % Na in amalgam leaving the electrolyzer minus % Na in amalgam entering. The basis is 96,500 A-sec/g-equivalent, 95.2% current efficiency, 23 g Na/g-equivalent, 454 g/lb, and 60 sec/min.

Figure 1-3 permits ready solution of the equation.

Typical Example

When the cell current is 100 kA and the percentages of sodium entering and leaving the electrolyzer are 0.020 and 0.170, respectively, what is the rate of flow of mercury amalgam? Connect 100 on the I-scale and 0.170—0.020, or 0.150, on the ΔNa-scale with a straight line. Read the flow rate of mercury amalgam on the R-scale as 2000 lb/min or 18 gal/min (on the basis of 111 lb amalgam/gal).

Note: Normally, the largest potential source of error is failure to obtain representative samples of amalgam.

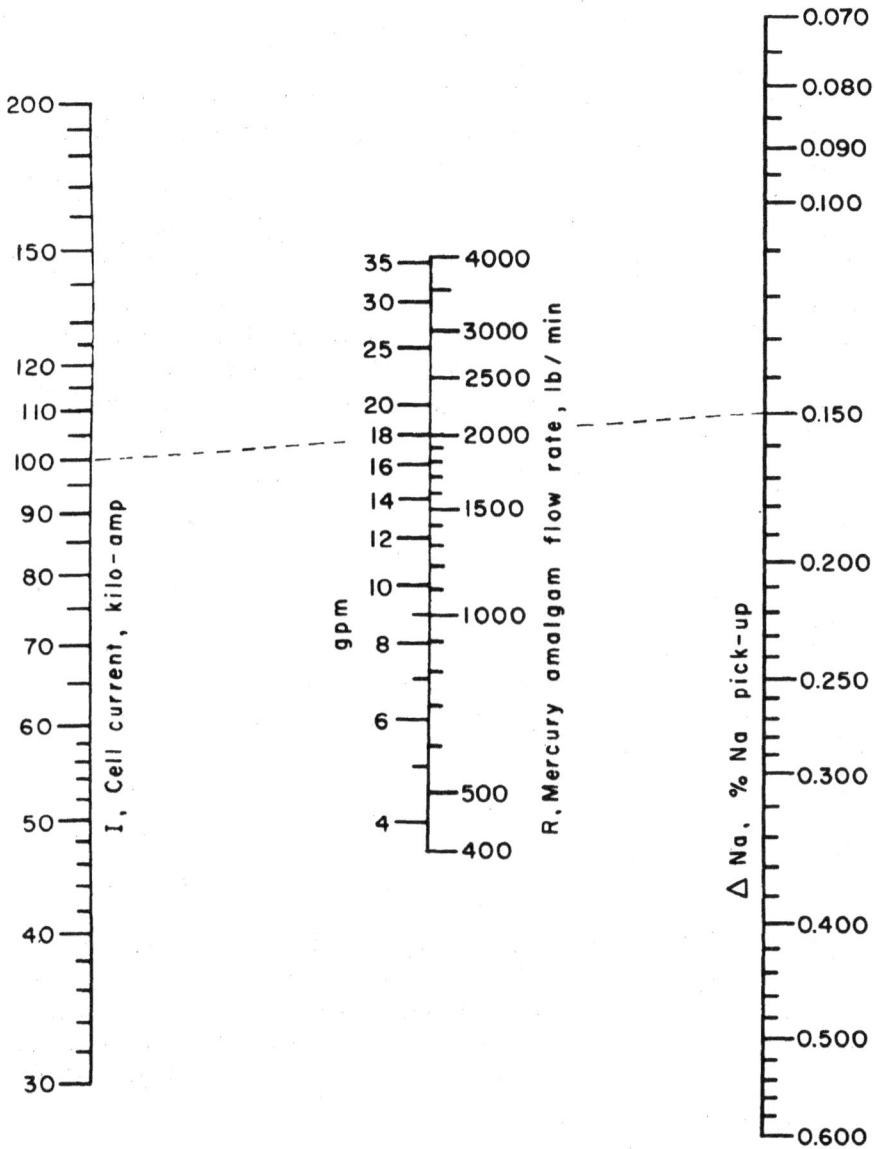

Figure 1-3

1-4   Viscous Flow Problems

J. D. HOSKINSON

Figure 1-4 was constructed for use as a guide in selecting pipe sizes in pumping or transfer of viscous resins common to urethane-foam industry.

In practice turbulent flow of viscous liquids (e.g. 10,000 centipoises) is virtually unattainable. For example, a flow of over 10,000 gal/min would be required through a 2-in. pipe, of a liquid at 10,000 centipoises and specific gravity of 1, to develop a Reynolds number of 2000. Since roughness of pipe has practically no effect on true viscous flow (except as roughness reduces effective cross-sectional area), Poiseuille's formula is applicable to the flow problems for which this chart is intended.

In modified form this formula may be stated as

$$P = \frac{27.3\,\mu\,G}{D^4}$$

where G = rate of flow, gal/min
        D = pipe diam, in.
        μ = viscosity, poises
        P = pressure drop, lb/in^2/1000 ft of pipe

Typical Example

Find the pressure drop through 1000 ft of 4-in. diameter pipe of a liquid having viscosity of 150 poises and flowing at 20 gal/min. Connect 150 on the μ-scale with 20 on the G-scale. Mark the intersection of this line on the reference line. Connect this point with 4 on the D-scale. This line intersects the P-scale at a pressure drop of 319 lb/in.^{2} for 1000 ft of pipe.

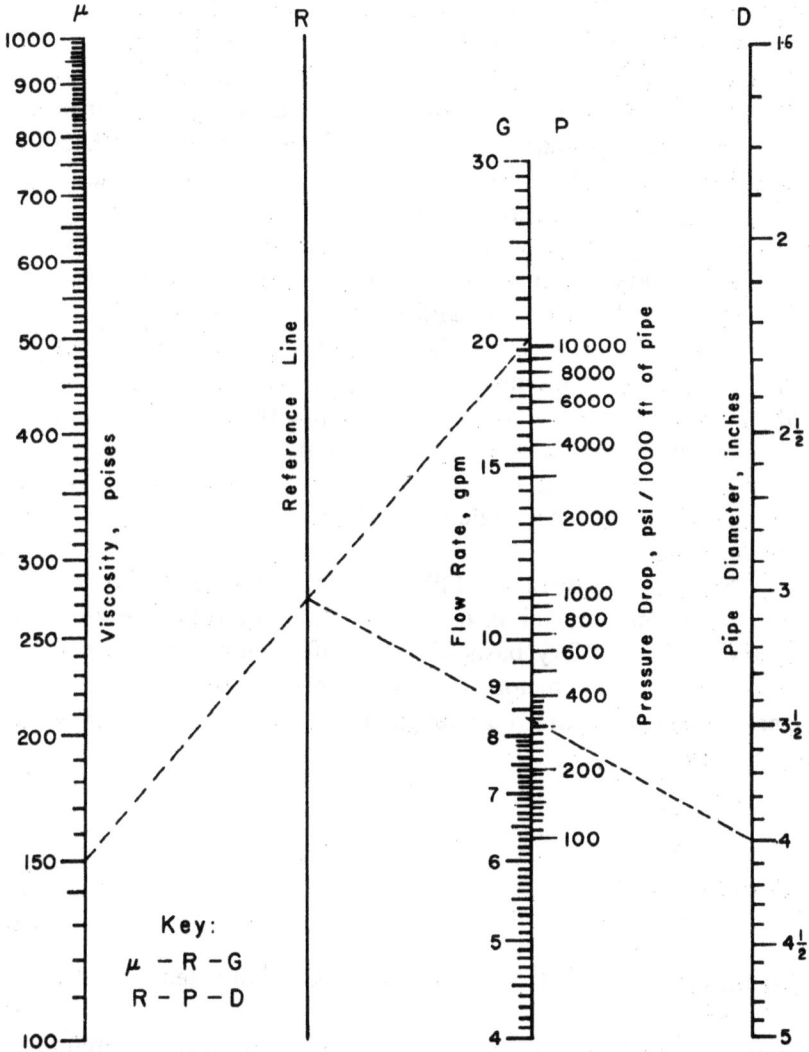

Figure 1-4

1-5   Orifice Sizing For Steam

R. L. PATTON

Figure 1-5 provides a quick method of solving flow equations for orifice meters handling steam.[1]

Obviously the nomograph is not suitable for precise calculations; for this reason it does not include such factors as corrections for thermal expansion of the primary device or Reynolds number. (These are usually negligible and seldom exceed a total of 1 % correction.) However, it is particularly useful for the following:

1) Spotting gross errors in precise calculations.

2) Preliminary determination of the feasibility of metering by the orifice method when line size and flow rates are known.

3) Deciding whether flange taps or pipe taps should be employed.

4) Choosing a manometer range for a given installation.

5) Obtaining rough figures for new capacity when a change in service is contemplated for a given installation.

6) Deciding whether to change the manometer range or to bore a new orifice plate in cases where an existing installation is over-ranged or under-ranged.

The "wet basis" side of the differential scale is used where a straight-wall mercury manometer is used with water or other liquid on the mercury surface. "Dry basis" is used when employing a spring (or spring-balance bell) meter, a ring-balance meter, a force-balance meter, or a meter with an air or gas purge to keep liquid away from the mercury.

Typical Example

Steam conditions: 100 lb/in.^{2} abs., 400°F total temperature

Line size: 6-in. ID

Meter: Mercury manometer, steam condensate on mercury surface, 200-in. range.

Maximum flow quantity: 30,000 lb/hr

The nomographic solution, shown by dotted lines, indicates that the orifice-diameter ratio is 0.65 for pipe taps and 0.73 for flange taps.

[1]Spink, L. K., *Principles and Practice of Flow Meter Engineering*, 7th ed., Foxboro Co., Foxboro, Mass, 1949.

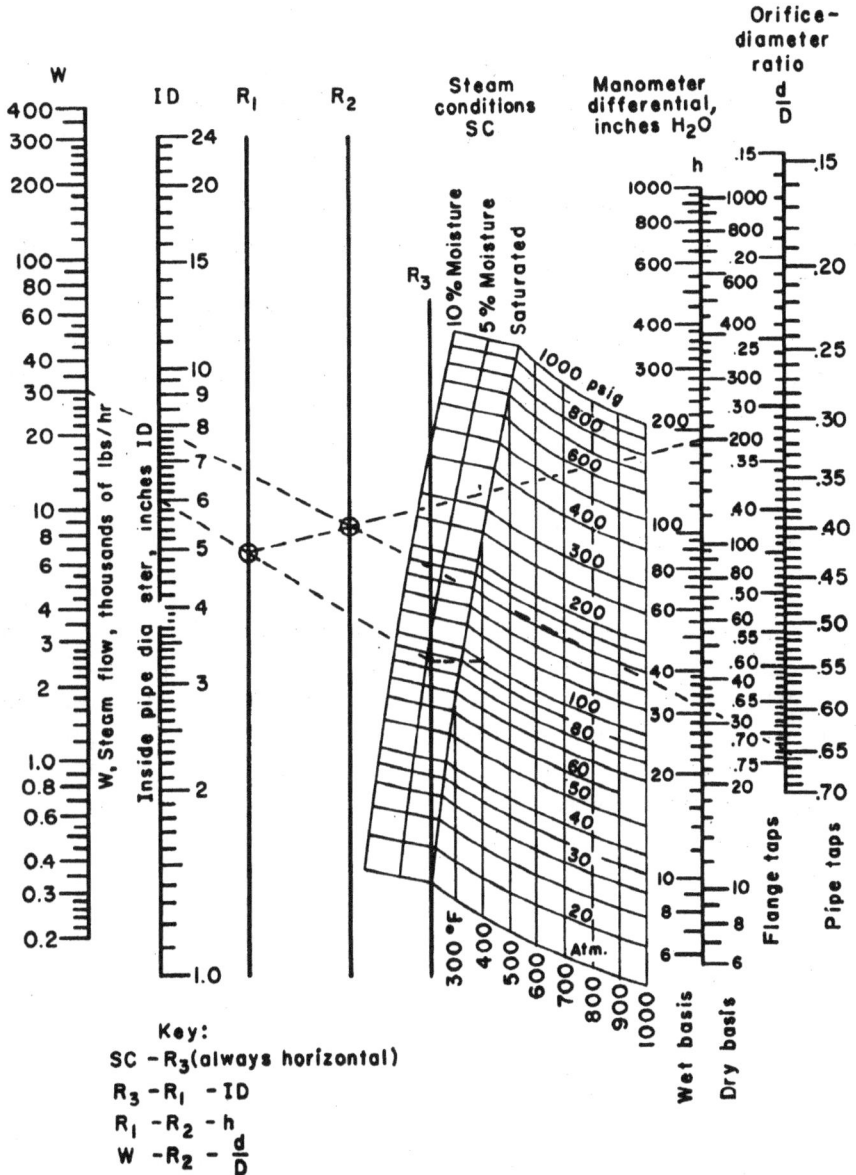

Figure 1-5

1-6   Pipe Size for Flashing Steam-Liquid Mixtures

ELIZABETH SHROFF

Normal methods of estimating sizes for pipe carrying flashing liquids are tedious and complicated. Figure 1-6 provides a quick way to determine the size of pipe carrying flashing steam-liquid mixture from steam traps.

Systems with multiple-steam-trap installations are handled by adding the orifice areas of all the traps and estimating the header size by using the total area.

For orifice areas greater than 0.1 in.², divide the area by some convenient number and determine the area of pipe required from the nomograph. The chart is also applicable for sizing of pipe carrying flashing mixtures from other blowdown or drain systems.

Typical Example

The example on the chart is for a system with 500 ft (equivalent length) of discharge pipe from a steam trap having an area of orifice of 0.021 in.², with a recommended safety factor of 3 and a back pressure of 10% of trap inlet pressure. Proceed from left to right. The return header size is between 1 and 1¼ in., so use the larger size 1¼ in. diameter pipe.

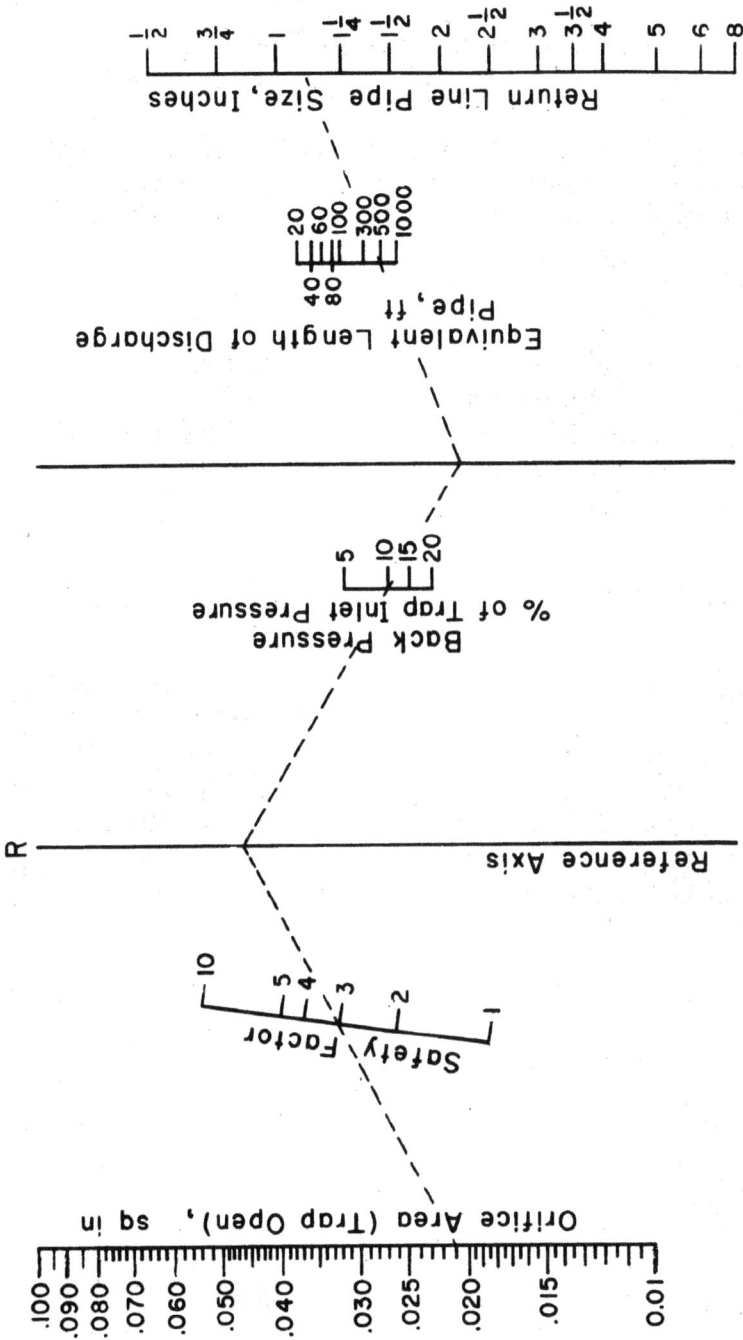

Figure 1-6

1-7   Sizing Steam Lines

J. H. BELL and D. S. DAVIS

Figure 1-7 aids one in sizing saturated steam and exhaust-transmission lines in refineries and steam plants. Constructed by means of well-known methods, the chart is based on the relationship:

$$\log W = c + \frac{d}{b} \left[\log (V-2) - a \right]$$

where W = rate of flow of saturated steam, lb/hr (1000)
       V = velocity of steam, (ft/min) (1000)
       a and b functions of D, nominal pipe diameters, in.
       c and d functions of P, steam pressure, lb/in.^{2}

Typical Example

Use of the chart is illustrated as follows: What is the nominal diameter of pipe that should carry 2000 lb/hr of saturated steam at a pressure of 50 lb/in.^{2} gage and velocity of 7200 ft/min? Connect 2 on the W-scale and 50 on the P-scale with a straight line, extending it to the R-axis. Draw a straight line from this intersection on the R-axis to 7.2 on the V-scale. Note that this line cuts the D-scale at a nominal pipe diameter of 3 in.

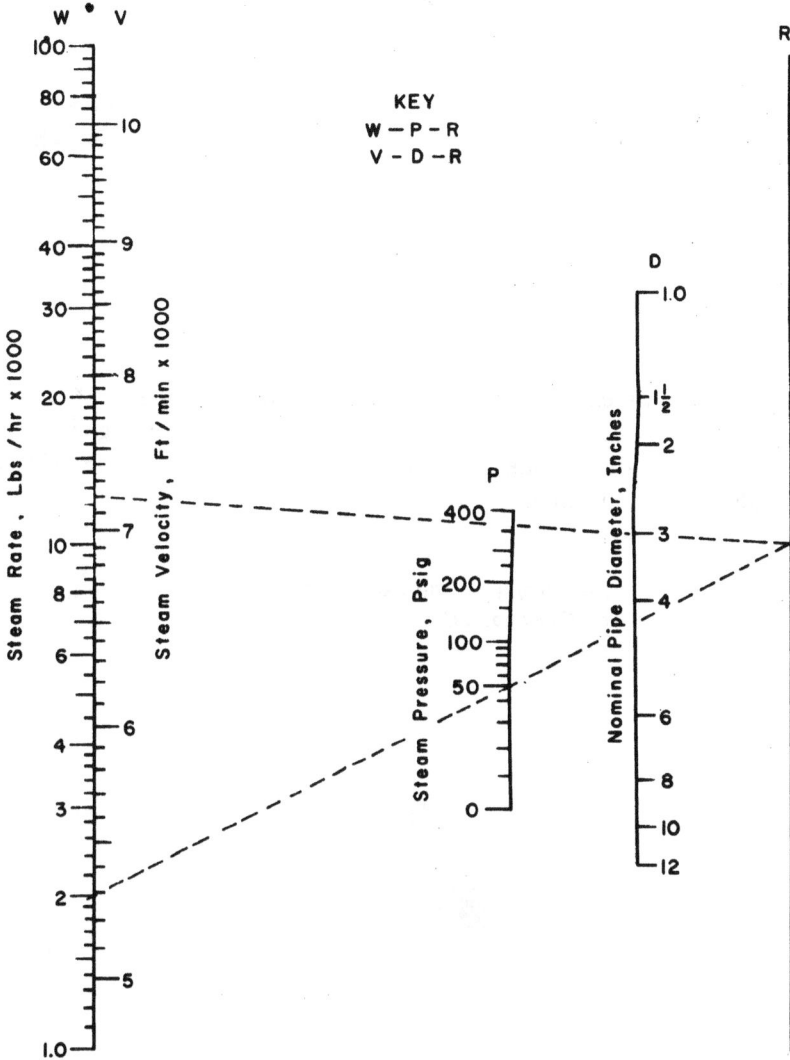

KEY
W — P — R
V — D — R

W ● **V**

Steam Rate, Lbs / hr x 1000

Steam Velocity, Ft / min x 1000

P

Steam Pressure, Psig

D

Nominal Pipe Diameter, Inches

R

Figure 1-7

1-8   Pressure Drops for Steam Through Valves and Fittings

D. S. DAVIS

When saturated steam (50 lb/in.^{2} gage) flows through valves and fittings, one can estimate pressure drops by use of Figure 1-8, which is based on reliable data[1] and constructed through regular methods.

Typical Example

The broken index line shows that when steam (50 lb/in.^{2} gage, saturated) flows through a 6-in. line, 250-lb cast iron, flanged, conventional 90° elbow at rate of 14,000 ft/min, a pressure drop of 0.4 lb/in.^{2} should be expected.

[1]"Flow of Fluids Through Valves, Fittings, and Pipe," pp 2-5, Technical Paper No. 410, Crane Co., Chicago, 1957.

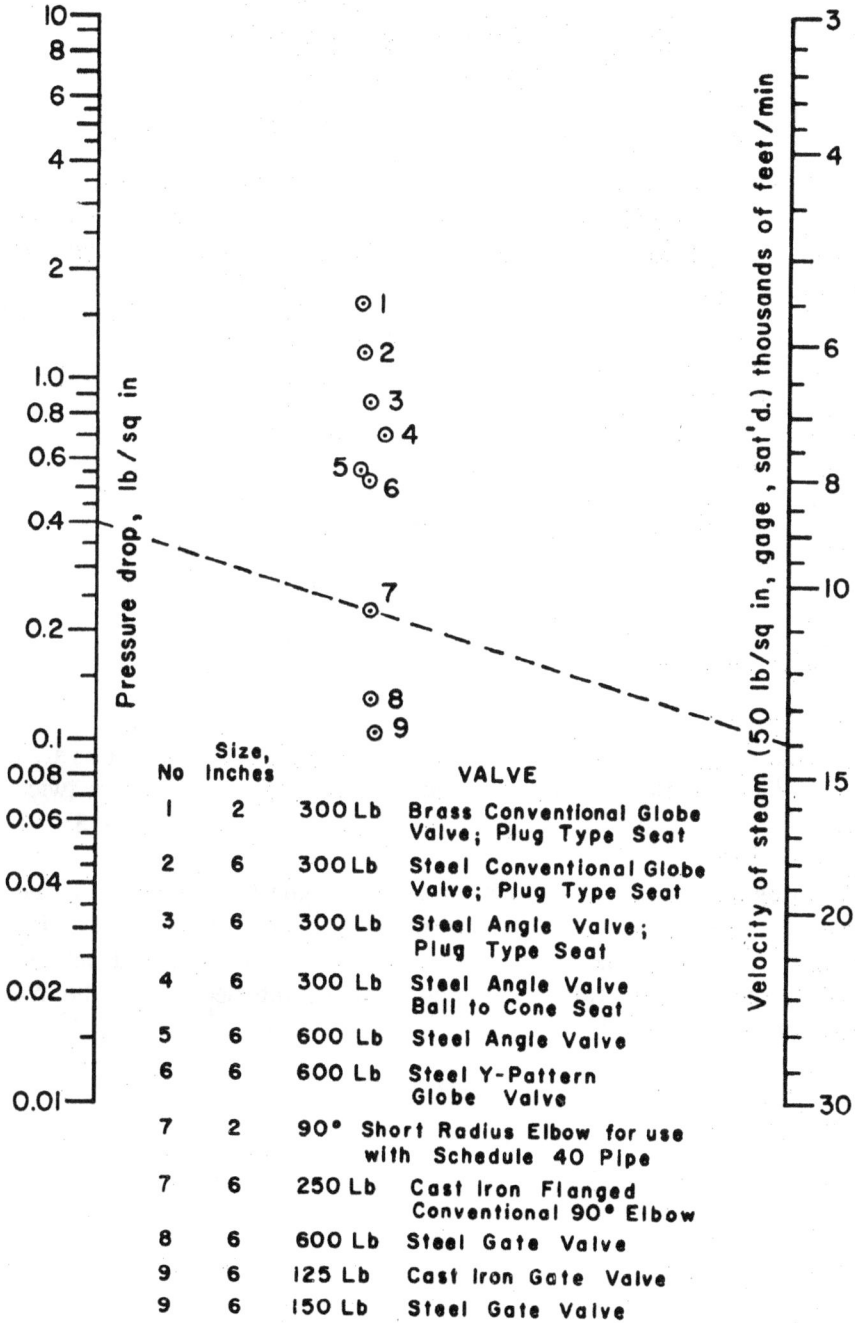

No	Size, Inches		VALVE
1	2	300 Lb	Brass Conventional Globe Valve; Plug Type Seat
2	6	300 Lb	Steel Conventional Globe Valve; Plug Type Seat
3	6	300 Lb	Steel Angle Valve; Plug Type Seat
4	6	300 Lb	Steel Angle Valve Ball to Cone Seat
5	6	600 Lb	Steel Angle Valve
6	6	600 Lb	Steel Y-Pattern Globe Valve
7	2		90° Short Radius Elbow for use with Schedule 40 Pipe
7	6	250 Lb	Cast Iron Flanged Conventional 90° Elbow
8	6	600 Lb	Steel Gate Valve
9	6	125 Lb	Cast Iron Gate Valve
9	6	150 Lb	Steel Gate Valve

Figure 1-8

1-9   Flow of Air Through Rounded, Circular Orifices

RICHARD K. REID

Figure 1-9 provides a quick determination of the rate of flow of air through a rounded circular orifice when the up-stream pressure is more than twice the down-stream pressure. The chart is based on the empirical equation

$$W = \frac{0.532\,AP}{T^{0.5}}$$

where W = rate of flow of air, lb/sec
        A = area of orifice, in.^{2}
        P = upstream pressure, lb/in.^{2} abs.
        T = upstream temperature, °R

Typical Example

What is the rate of flow of air through a rounded orifice with an area of 3 in.^{2} when the upstream pressure, 200 lb/in.^{2}, is more than twice the downstream pressure—and the upstream temperature is 580°R (or 120°F)?

Following the broken index lines on the chart, connect 3 on the A-scale and 200 on the P-scale with a straight line, and note the intersection with a, the turning axis. Connect this point and 580 on the T-scale with a straight line, and read the intersection on the W-scale as 13.3 lb/sec of air.

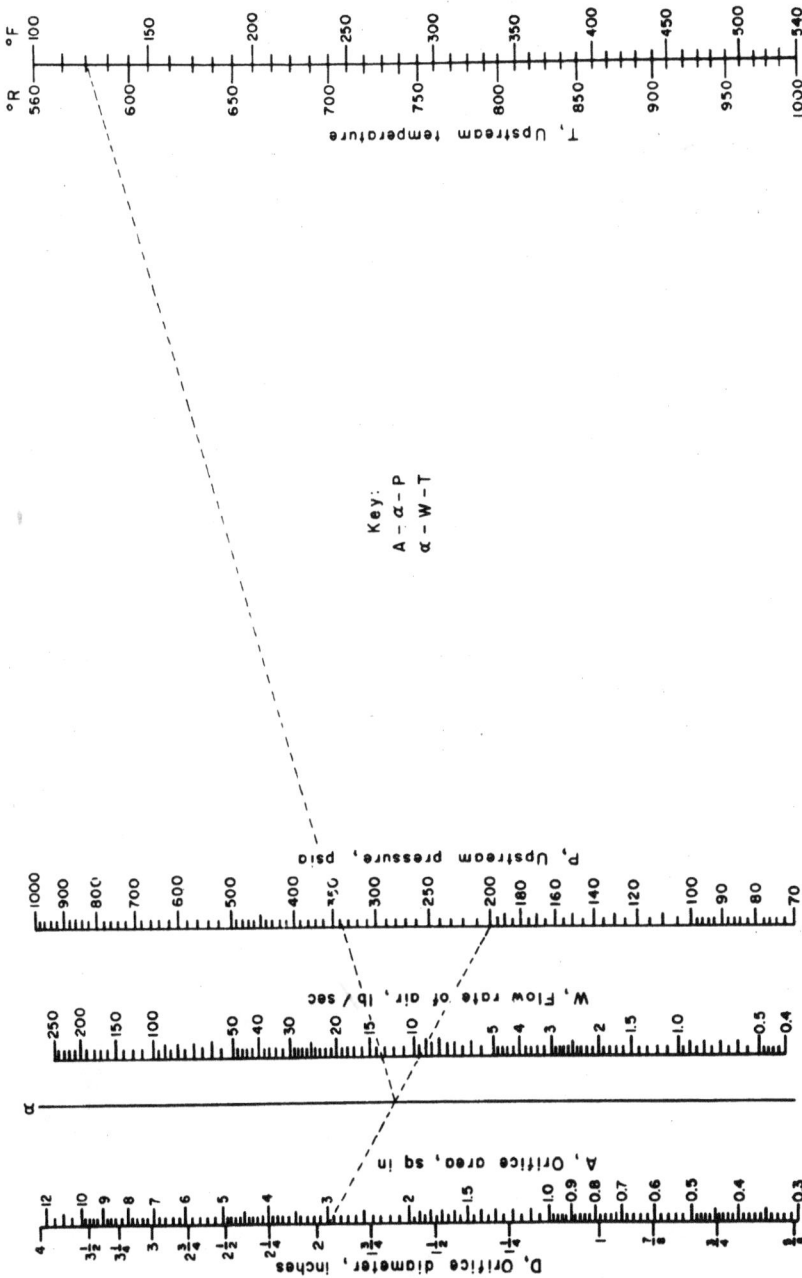

Figure 1-9

1-10   **Estimating Air Infiltration or Ventilation Quantities**

F. CAPLAN

Air-change computations are commonly made for estimating the amount of air infiltrating into rooms and for estimating air quantities required for ventilation. A nomographic solution is presented in Figure 1-10, in accord with the equation

$$R = 16.67NV$$

where   R   =   air rate, ft^3/min
         N   =   no. of air changes per hr
         V   =   volume of room, 1000 ft^3/hr

Typical Example

For what air rate, R ft^3/min, should a fan be sized to provide 4 air changes per hr in a room with a volume of 21,000 ft^3?

Align N = 4 with V = 21 and read R = 1400 ft^3/min.

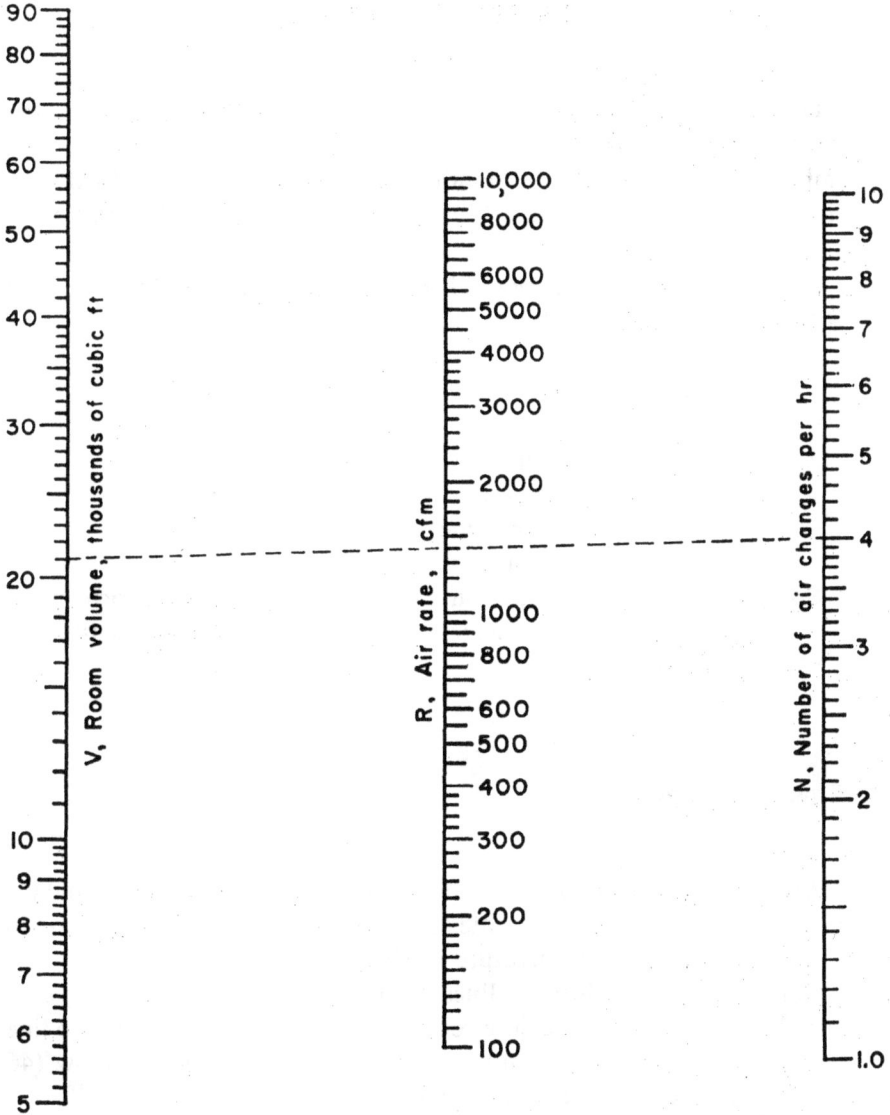

Figure 1-10

1-11   Temperature and Pressure Corrections to Rotameters Calibrated for Gases at Standard Conditions

F. E. SCHRAGE

Often rotameters supplied for gas-flow measurement are calibrated for a particular gas flowing at standard conditions. In actual use, ambient conditions may be significantly different from standard conditions. In such cases correction must be applied to rotameter reading to obtain true flow.

Assuming perfect gas laws apply, following equation may be used to correct rotameter reading:

$$Q = R \sqrt{\frac{(P+14.7)(273)}{(14.7)\,(t+273)}}$$

where Q = actual gas flow, wt/vol
        R = rotameter reading
        P = pressure in rotameter, lb/in.^{2} gage
        t = temperature in rotameter, °C

Figure 1-11 solves this equation. Units of actual gas flow correspond to the units for which rotameter is calibrated. The decimal point is similarly related.

Typical Example

Air is flowing through a rotameter calibrated for air at standard conditions. The meter indicates flow of 40 ft^3/min. Pressure is 10 lb/in.^{2} gage and the temperature is 80°C.

Connect proper values on lines A and B. Read the gas flow, corrected to atmospheric pressure, on line C. Connect this value on line C with the proper temperature on line D. Read the gas flow (46 ft^3/min), corrected to standard conditions, on line E.

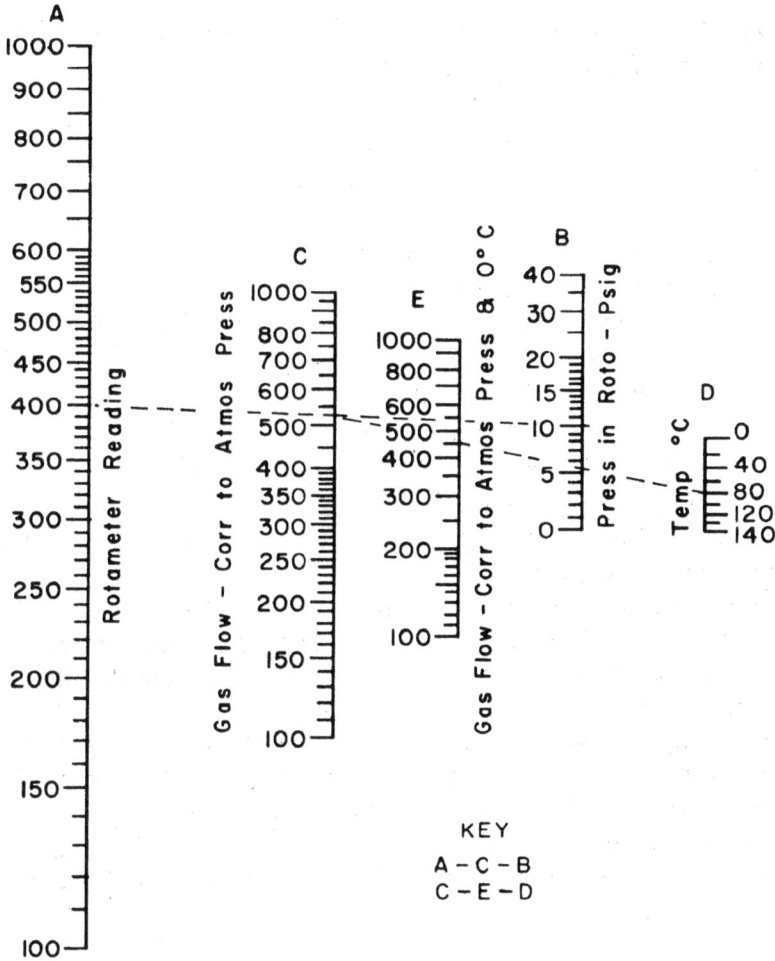

Figure 1-11

1-12   Effective Orifice Area for Pressure Relief Device

R. D. BIGGS

Figure 1-12 gives the effective orifice area of a pressure relief device for discharge of a vapor according to the equation

$$A = \frac{Q}{306 \, \lambda \, P \sqrt{\dfrac{M}{T+460}}}$$

where   A   =   effective orifice area, in.^{2}
        Q   =   rate of heat input, Btu/hr
        λ =   latent heat of vaporization of stored liquid, Btu/lb
        P   =   absolute pressure of vapor at inlet conditions, lb/in.^{2} abs.
        M   =   molecular weight of vapor
        T   =   temperature of vapor at inlet conditions, °F

This equation, the API-ASME Code formula, is applicable to selection of pressure relief devices designed to protect tanks containing volatile liquids.

The value of Q depends upon exposed surface area of tank and an arbitrary choice of value of heat flux. Reference No. 1 contains a review of the various methods which have been recommended for calculating Q.[1]

The formula is based on the adiabatic, reversible expansion of an ideal gas through a well designed pressure relief device (coefficient of discharge = 0.97) where a sufficient pressure differential is available to insure critical flow.

The orifice area so calculated will be conservative for real gases.

Typical Example

To solve the equation, connect the value of the temperature with the molecular weight to intersect the reference scale A.

Connect this point on scale A with the value of absolute pressure to intersect the reference scale B. Connect this point with the value of latent heat to intersect the reference scale C.

Connect this point with the value of heat absorption rate to obtain the value of the effective orifice area required.

[1]Sylvander, N. E., and Katz, D. L., "Design and Construction of Pressure Relieving Systems," *Petroleum Processing*, July, August, September, 1948.

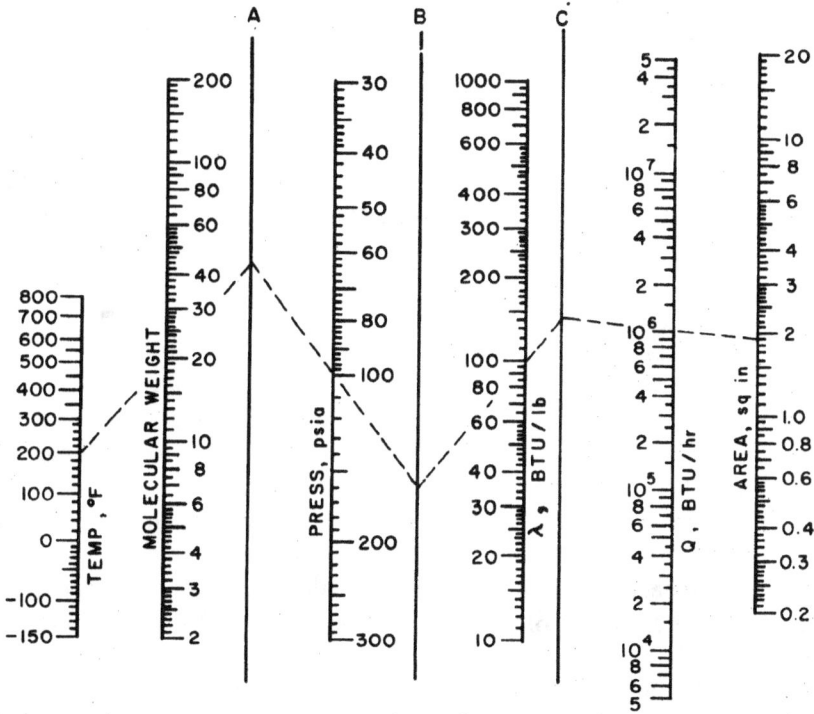

Figure 1-12

1-13   Fluidizing Gas Velocities

JOHN D. GABOR

In operating fluidized-bed columns, it is necessary to know the range of gas velocities in which fluidization will occur for a particular system. Figure 1-13 enables quick estimates to be made of both the minimum-fluidization and terminal velocities. (Latter quantity is that velocity at which the fluidized particles would be carried out of the column.)

Typical Example

It is desired to know the range of gas velocities in which 0.4-mm diameter copper shot can be fluidized by air at room temperature and atmospheric pressure. Other physical properties given for this system are:

gas density $= 0.0012$ g/cm³
gas viscosity $= 0.018$ centipoise
particle density $= 8.9$ g/cm³

Connect 0.0012 on the ρ_f-scale with 0.018 on the μ-scale. Extend this line to intersect both the A- and E-scales. Align the intersection point on the A-axis with approximately 8.9 (ρ_s-ρ_f) on the left side of the $\Delta\rho$-scale. Mark the intersection the resulting line makes with the B-scale.

Now connect the point on the B-scale with 0.4 on the d-scale. Mark the intersection of the resulting line on the R_1-axis. Following the slope of the neighboring pattern lines, draw an intermediate line from this point to the R_2-scale.

Following the key listed with the nomograph, align the point on R_2 with the previously marked point on the E-axis and extend the resulting line to D. Drawing a line from 0.4 on the d-axis through the mark on the D-scale, find the minimum fluidization velocity to be 1.45 ft/sec on the V-scale.

The terminal velocity is found in the same manner, except that the right side of the $\Delta\rho$-scale is used. Following the same procedure, the terminal velocity is found to be 23 ft/sec. Thus, it is estimated that the range of gas fluidizing velocities for the system is between 1.45 and 23 ft/sec.

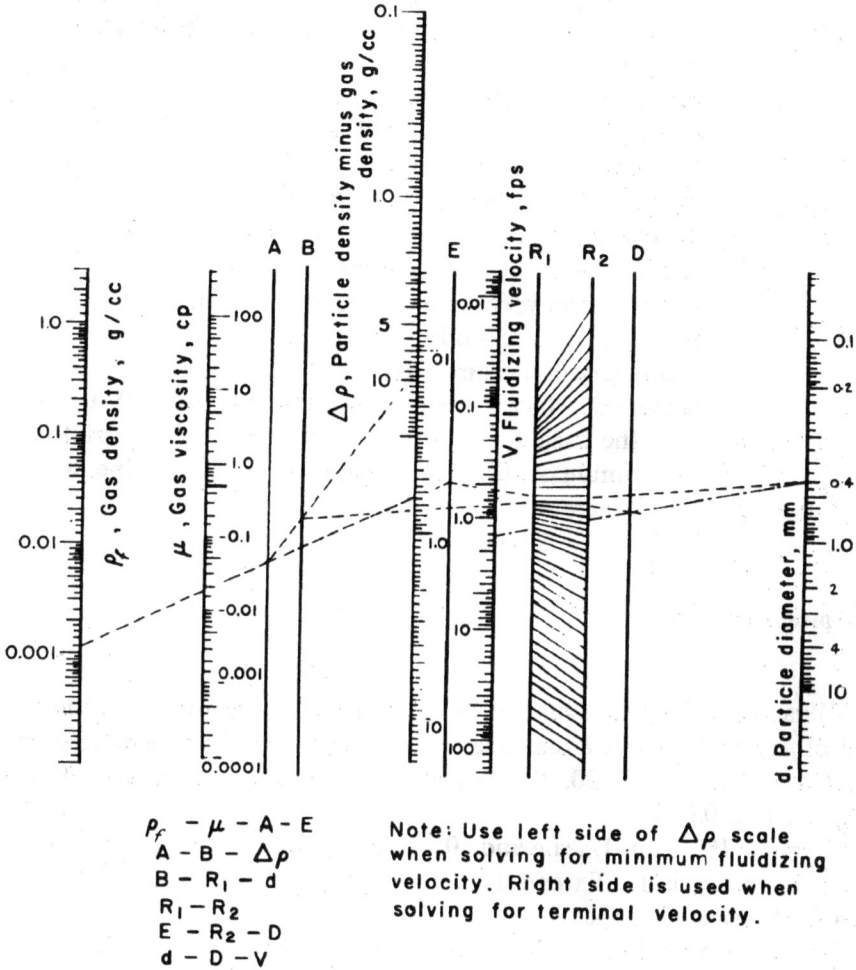

$\rho_f - \mu - A - E$
$A - B - \Delta\rho$
$B - R_1 - d$
$R_1 - R_2$
$E - R_2 - D$
$d - D - V$

Note: Use left side of $\Delta\rho$ scale when solving for minimum fluidizing velocity. Right side is used when solving for terminal velocity.

Figure 1-13

1-14 Transmission of Gas at High Pressure

F. P. VANCE

The following equation[1] is useful in design of pipe lines for transmission of gas at high pressures:

$$Q=\frac{80.8}{\rho_0\mu^{0.0814}}\left[\frac{D^{4.85}M(P_1^2-P_2^2)}{LZ_{av}T}\right]^{0.541}E$$

where Q = quantity of gas transported in ft³/day measured at P_0 and T_0

   μ = absolute viscosity of gas, lb/ft-sec

   ρ_0 = density of gas, lb/ft³ at P_0 and T_0

   D = inside diameter of pipe, in

   M = molecular weight of gas, lb/mole

   P_1 = discharge pressure from compressors, lb/in.² abs.

   P_2 = suction pressure to compressors, psia

   L = distance between compressor stations, miles

   Z_{av}= average compressibility factor

   T = flowing temperature, °R

   E = experience factor for adjustment of formula, 0.90

This equation is the basis for Figure 1-14, which facilitates solution of the case for optimum spacing of compressor stations, pipe sizes, and other pertinent factors.

Typical Example

What should be the diameter of a pipe needed to transmit 10 million ft³ of gas per day over a distance of 15 miles at 100°F, if the molecular weight of the gas is 20, the viscosity is 0.0001 lb/ft-sec, and $(P_1^2-P_2^2)/Z_{av}$ is 150,000?

Connect 100°F on T-scale and 10^{-4} on the μ-scale to locate an intersection with axis I. Extend a line from this intersection through 20 on M to meet the K-axis.

Connect 15 on L and $1.5(10)^5$ on the scale for $(P_1^2-P_2^2)/Z_{av}$ and mark the intersection with axis III. Extend a line from this intersection through 10^7 on the Q-scale to meet axis II.

Extend the line from intersection found on the K-axis through intersection on axis II to meet the D-scale in the desired value of 6.7 in.

---

[1]"High Pressure Pipe Line Research," Project 49, Clark Bros. Co., Inc., Olean, N.Y., 1942.

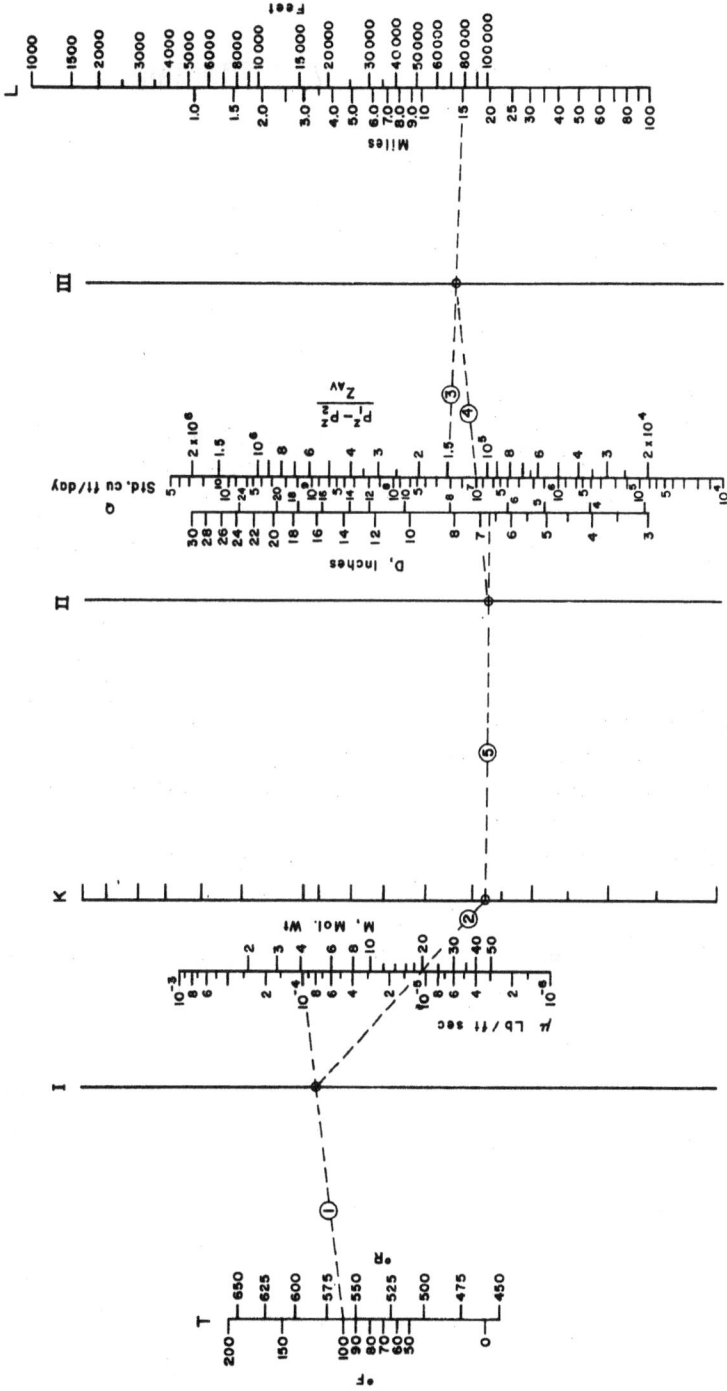

Figure 1-14

1-15   Flow of Water from Full Horizontal
Open-end Pipes

CLIFFORD L. DUCKWORTH

Figure 1-15 presents a means of determining the rate of flow of water from full horizontal open-end pipes. It is based on the equation[1]

$$q = 1.04AD$$

where A = internal transverse area of the pipe, $in.^2$
         D = horizontal distance from pipe opening to a point where the water stream has fallen one foot, in.
         q = rate of flow of water, gal/min.

Typical Example

Water is pumped out of a full horizontal pipe of 4-in. nominal diameter; the stream drops one foot at a point 75 in. from pipe opening.
Connect the point of the d-scale for 4-in. nominal diameter and D = 75 with a straight-edge.
Read the rate of flow of water as 990 g/min on the q-scale.

[1]Brooke, M., *Chem. Eng.*, **64** (1) 264, 1957.

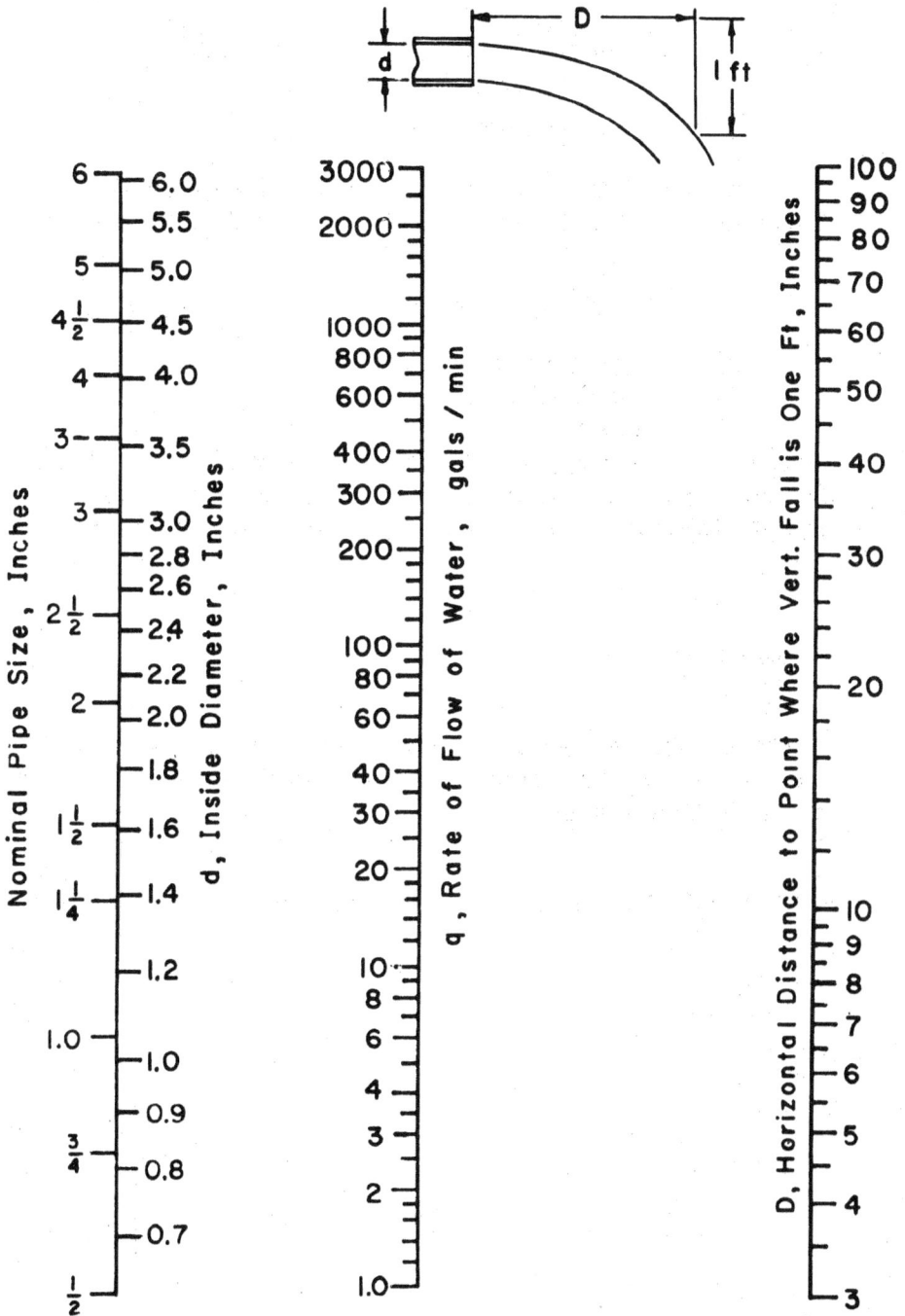

Figure 1-15

1-16   Flow of Water from a Vertical Open-end Pipe

CLIFFORD L. DUCKWORTH

Figure 1-16 makes it possible to determine the quantity of water flowing from a vertical open-end pipe. It is based on the equations[1]

$$V = 5.84 \, D_1^{2.025} \, H_1^{0.53}$$
$$V = 8.8 \, D_2^{1.29} \, H_2^{1.24}$$

where V =   volume, ft³/sec
         D =   pipe diameter, ft
         H =   head, ft.

The first equation (D_1 and H_1) is for jet flow; second (D_2 and H_2) is for weir flow. For greater practicality, units of measure have been changed to gal/min, nominal pipe size, and in., respectively.

Typical Example

Water is flowing from a 6-in. pipe. Distance from pipe opening to crest is 9 in. The line connecting the D_1 and H_1 scales (for jet flow), crosses V at 560 gal/min.

[1]Brook, M., *Chemical Engineering*, **64** (2) 290, 1957.

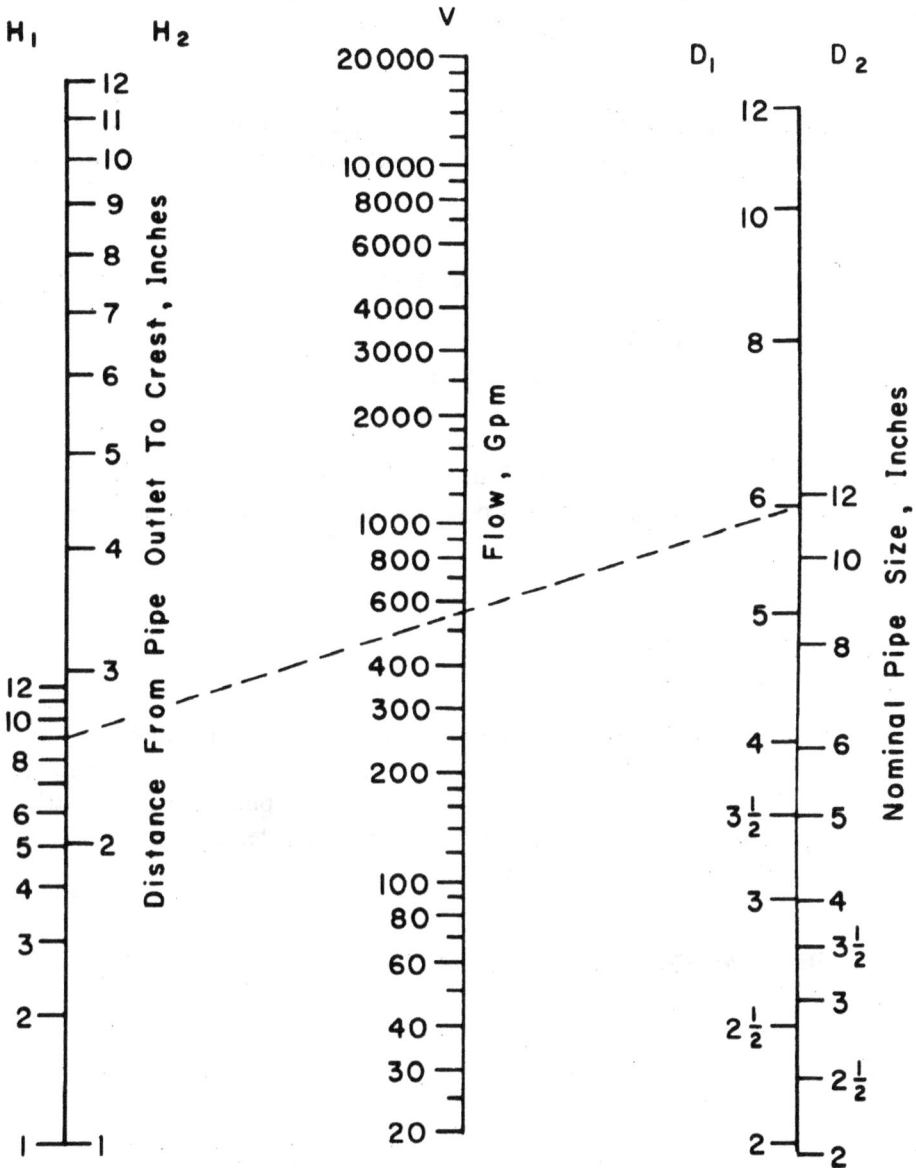

H₁ **H₂** **V** **D₁** **D₂**

Figure 1-16

1-17   Flow of Water Through Concrete Pipes

CLIFFORD L. DUCKWORTH

Figure 1-17 permits one to estimate pressure drop in water flowing through concrete pipes of varying construction. It is based on the equation[1]

$$H=\frac{V^2}{C^2\,d^{1.25}}$$

where   V = velocity in ft/sec
       d = internal pipe diameter
      C = a factor correlating type of construction
      H = pressure drop in feet of water

It is applicable to pipes from 8 to 63.5 in. in diameter. For expansion, flow velocity and pressure drop have been drawn as double scales.

Typical Example

Water is flowing through a 16 in. pipe laid without removal of the mortar squeeze. The water velocity is 16 ft/sec. A line is drawn between 16 on the V-scale and the appropriate point on the C-scale. A line connecting 16 on the d-scale with the intersection on the reference line R crosses H at 110.

[1]Brook, M., *Chemical Engineering*, **64** (2), 290, 1957.

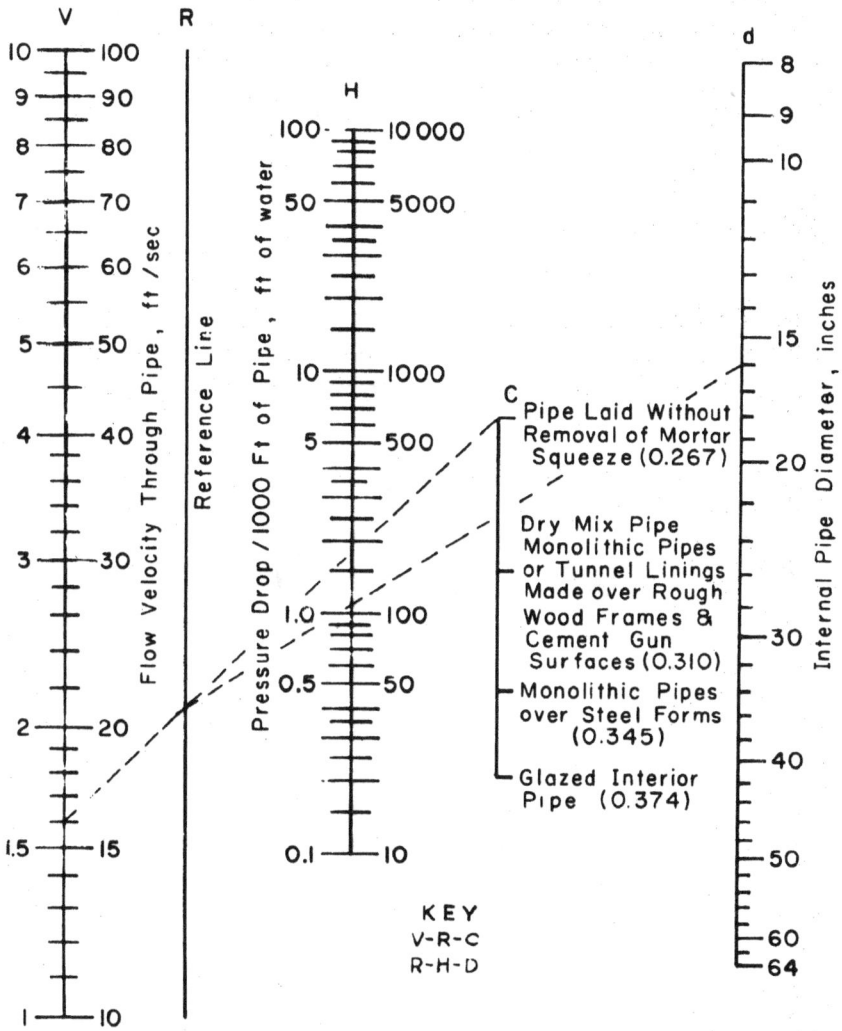

Figure 1-17

1-18   Pressure Drops for Water
Through Valves

D. S. DAVIS

Figure 1-18, based on reliable data,[1] and constructed by regular methods, enables one to estimate the pressure drops, when water flows through various size and type valves from this particular manufacturer.[1]

Typical Example

The broken index line shows that one should expect a pressure drop of 2 lb/in. when water flows through a ¾-in., 150-lb cast iron, Y-pattern, globe valve, with a flat seat, at a rate of 8.5 ft/sec.

[1]"Flow of Fluids through Valves, Fittings, and Pipe," pp 2-4, Technical Paper No. 410, Crane Co., Chicago, 1957.

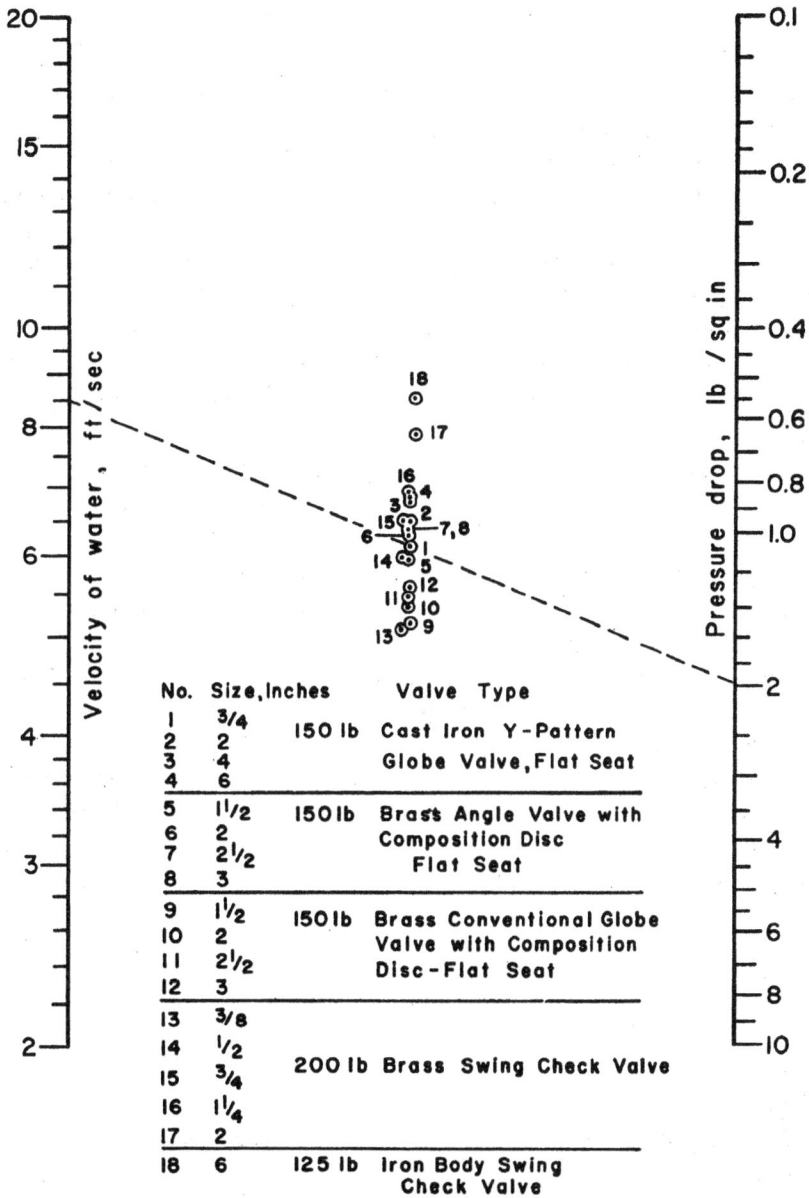

No.	Size, Inches	Valve Type
1	3/4	150 lb Cast Iron Y-Pattern
2	2	Globe Valve, Flat Seat
3	4	
4	6	
5	1 1/2	150 lb Brass Angle Valve with
6	2	Composition Disc
7	2 1/2	Flat Seat
8	3	
9	1 1/2	150 lb Brass Conventional Globe
10	2	Valve with Composition
11	2 1/2	Disc-Flat Seat
12	3	
13	3/8	
14	1/2	
15	3/4	200 lb Brass Swing Check Valve
16	1 1/4	
17	2	
18	6	125 lb Iron Body Swing Check Valve

Figure 1-18

1-19   Orifice Sizing for Liquid-Flow
Measurement

R. L. PATTON

Figure 1-19 provides a quick method of solving flow equations for orifice meters handling liquids.[1]

Not designed for precise calculations, the nomograph does not include such factors as corrections for thermal expansion of the primary device or for Reynolds number. (These are usually negligible and seldom exceed a total of 1 % correction.) However, it is particularly useful for the following:

1) Spotting gross errors in precise calculations.
2) Preliminary determination of the feasibility of metering by the orifice method when line size and flow rates are known.
3) Deciding whether flange taps or pipe taps should be employed.
4) Choosing a manometer range for a given installation.
5) Obtaining rough figures for new capacity when a change in service is contemplated for a given installation.
6) Deciding whether to change the manometer range or to bore a new orifice plate in cases where an existing installation is over-ranged or under-ranged.

The "wet-basis" side of the differential scale is needed where a straight-wall mercury manometer is used with water or other liquid on the mercury surface. "Dry basis" is used when employing a spring (or spring-balance bell) meter, ring-balance or force-balance meters, or a meter with an air or gas purge to keep liquid away from the mercury.

Typical Example

Liquid: water at 150°F
Line size (D): 4 in. ID
Meter: mercury manometer, water on mercury surface, 100 in. range
Maximum flow rate (V): 200 gal/min
Determine the orifice-diameter ratio for use with pipe taps and with flange taps.
Connect 4 in. on the D-scale with 0.98 (specific gravity of water at

150°F) on the G_f-scale. The intersection of this line with the A-axis is next connected with 1.0 (base specific gravity of water) on the G_b-scale. Now join the intersection of this line and the B-scale to 100 in. on the H-scale. A point has now been formed on the C-axis. Connect this point with 200 gal/min on the V-scale, and read an orifice-diameter ratio of 0.55 for pipe taps and 0.60 for flange taps.

[1]Spink, L. K., *Principles and Practice of Flow Meter Engineering*, 7th Ed., Foxboro Co., Foxboro, Mass. 1949.

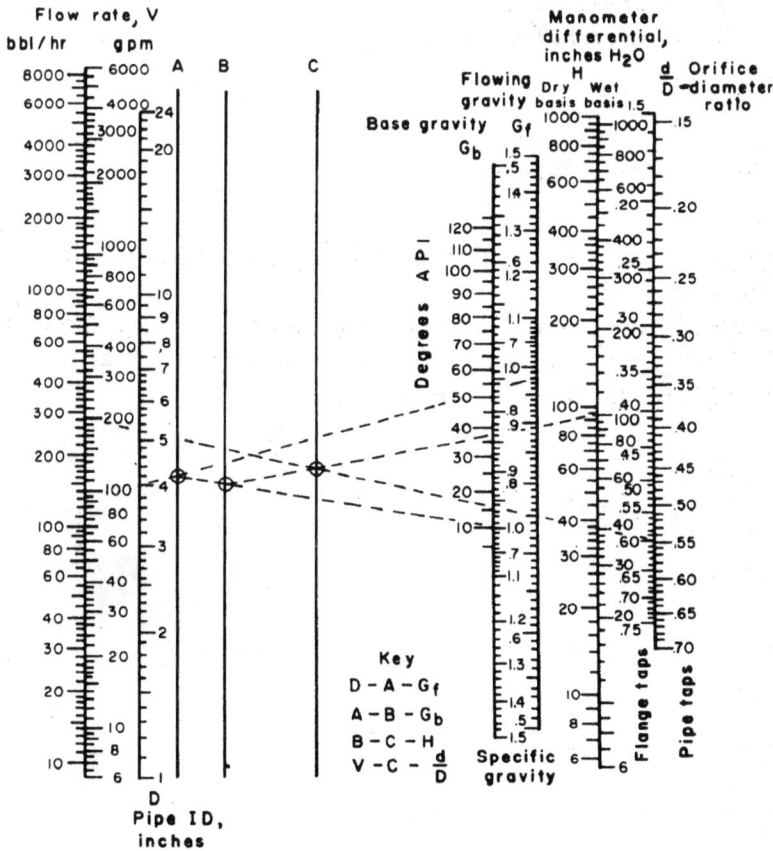

Figure 1-19

1-20   Hydrostatic Head of Liquids

S. J. SALVA and L. J. CRULL

Figure 1-20 is useful in computing the hydrostatic head of liquids that is due to static pressure, which is frequently necessary when specifying pumps or piping systems for various liquids.

The nomograph is based on the following equation:

$$H = \frac{144\,P}{W}$$

where   H = hydrostatic head, ft
        P = pressure, lb/in.^{2}
        W = weight/ft^3 of liquid

Typical Example

Determine the hydrostatic head of ammonia if the pressure to be used is 10 lb/in.^{2}

Extend a straight line from 10 on the P-scale to 0.89 (specific gravity of ammonia) on the ϱ-scale.   Read the head as 25.9 ft at the intersection of this line with the H-scale.

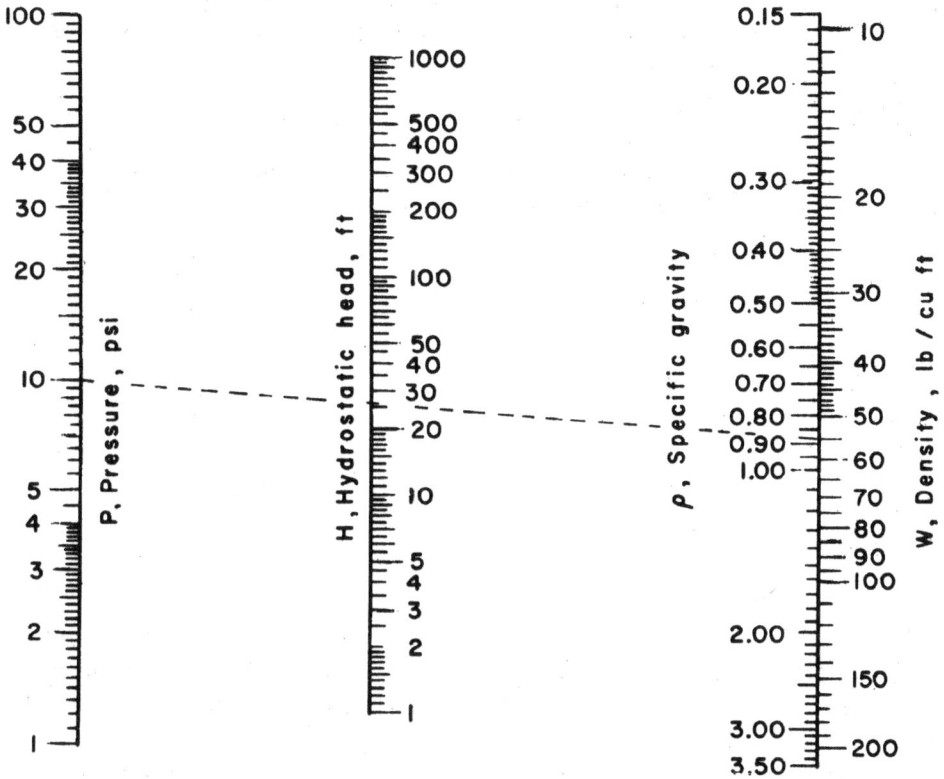

Figure 1-20

1-21   Equivalent Diameters of Tube Bundles

IRVING GRANET and D. S. DAVIS

Equivalent diameters of shell side of tube bundles, for use in heat-transfer and fluid-flow calculations, can be determined by means of Figure 1-21.

Equivalent diameter d_e is equal to four times the flow area, divided by the wetted perimeter. Kern[1] states that regardless of whether flow is cross or parallel equivalent diameter should be taken as if flow were parallel to tubes.

On this basis, for square and equilateral triangular pitch (Figure 1-21), pertinent expressions appear in the following table, where P is pitch and d_0 is the outer diameter of tube.

	Flow Area	*Wetted Perimeter*	*Equivalent Diameter*
Square	$P^2 - \dfrac{\pi d_0^2}{4}$	πd_0	$\dfrac{4P^2}{\pi d_0} - d_0$
Triangular	$\dfrac{\sqrt{3}\, P^2}{4} - \dfrac{\pi d_0^2}{8}$	$\dfrac{\pi d_0}{2}$	$\dfrac{2\sqrt{3}\, P^2}{\pi d_0} - d_0$

Typical Example

Use of nomograph Figure 1-21 is illustrated as follows: What is the equivalent diameter of the shell side of a tube bundle, where outer diameters of tubes are 1 in. and tubes are arranged on equilateral triangles so that the pitch is 1.8 in.?

Connect $P = 1.8$ and $d_0 = 1$ (on the scale for equilateral triangular pitch) with a straight line. Note that this line meets the d_e-scale at an equivalent diameter of 2.57 in.

[1] Kern, D. Q., *Process Heat Transfer*, McGraw-Hill Book Company, New York, 1950.

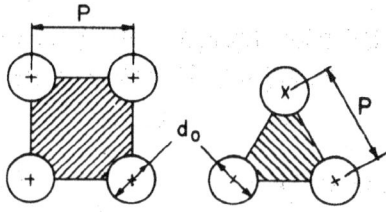

Figure I

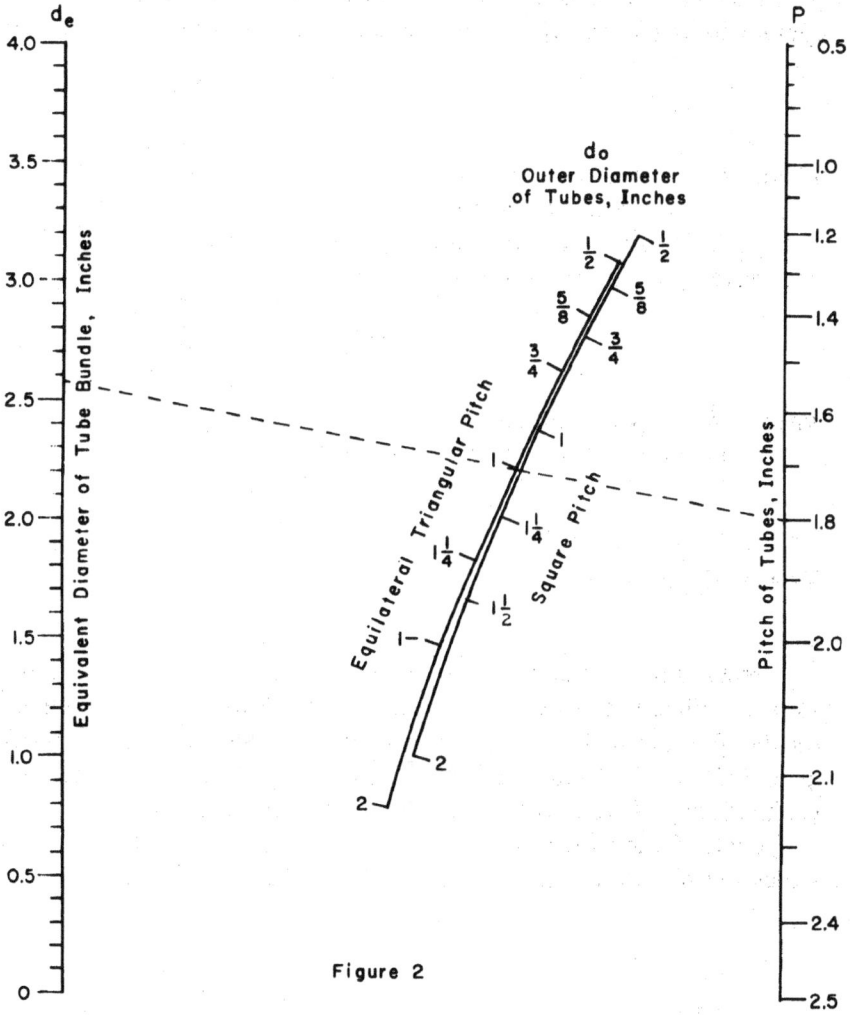

Figure 2

Figure 1-21

1-22   Hydraulic Radius for Rectangular
Sections Flowing Full

PAUL WHITE

Formulas relating to flow in open or closed conduits customarily express the size and shape by means of the hydraulic radius. For closed conduits, it is the distance obtained by dividing the area of the cross section of the liquid flowing in the conduit by the wetted perimeter of the conduit. The perimeter is the length of the line of contact between the liquid cross-section and the conduit wall. Thus,

$$R = \frac{A}{P}$$

where   R   = hydraulic radius, in.
        P   = wetted perimeter, in.
        A   = area of cross-section of liquid, in.^{2}
For rectangular sections flowing full,

$$R = \frac{WD}{2W + 2D}$$

where   W   = width of conduit, in.
        D   = depth of conduit, in.

Typical Example

A rectangular conduit is 7 in. wide and 5.3 in. deep (inside dimensions). What is the hydraulic radius when flowing full? Connect 7 on the W-scale with 5.3 on the D-scale. Intersection on the R-scale gives 1.5″ for hydraulic radius. This is close enough for hydraulic calculations. Actual value from the equation is 1.508.

To extend the range of Figure 1-22, move the decimal point for *all* scales the same number of places to the right or left as needed.

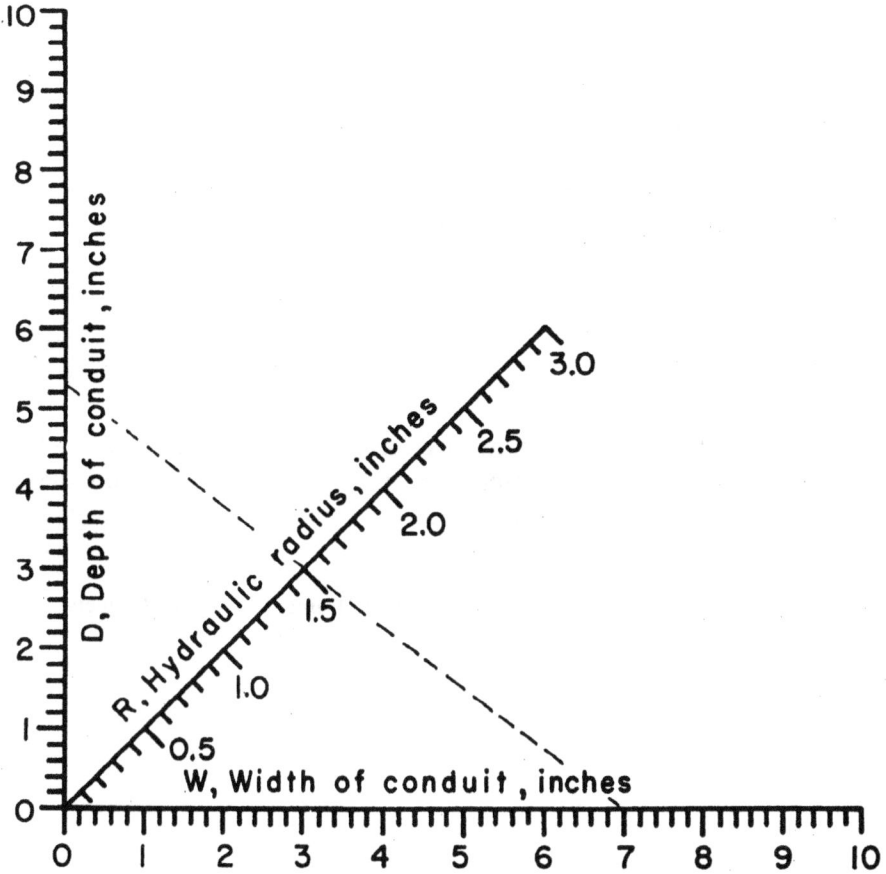

Figure 1-22

1-23   Flow of Water Over a Cipolletti Weir

S. E. HENRY

The flow of water over a sharp-edged Cipolletti weir — where the sides slope upward and outward — is described by the following formula[1]

$$Q = 3.030 \, lh^{1.5}$$

where Q = flow rate, gal/min
        l = length of crest, in.
        h = head of water measured from weir crest, in.
The equation can be solved readily by means of Figure 1-23.

Typical Example

How much water will flow over a 17.5 in. Cipolletti weir when the head is 3.5 in.? Connect 17.5 on the l-scale with 3.5 on the h-scale. Extend the line and read 350 gal/min. on the Q-scale.

[1]*Handbook of Water Control*, p. 45, Armco Drainage & Metal Products, Inc., 1949.

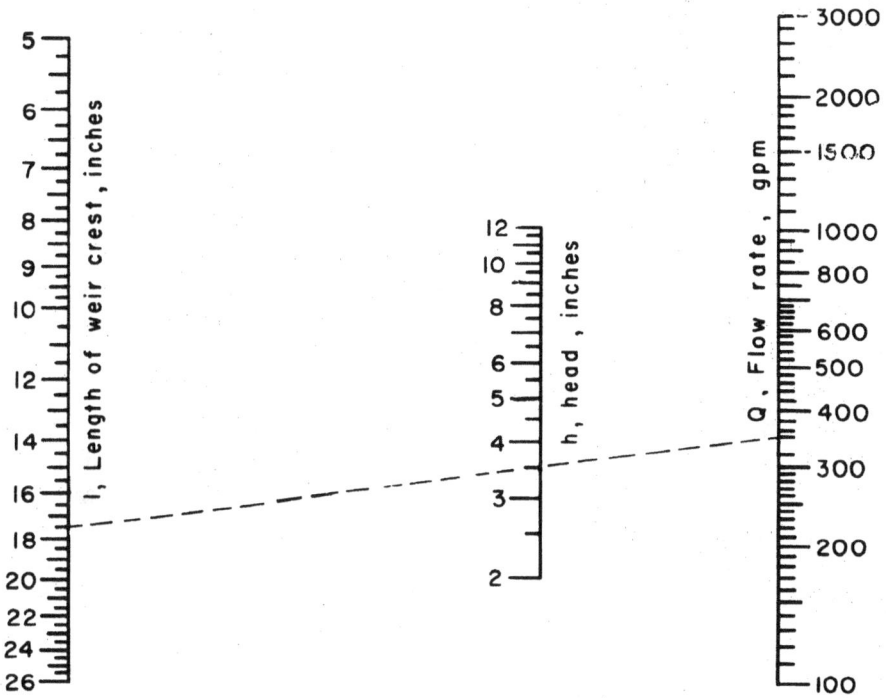

Figure 1-23

1-24   Rectangular and Triangular Weirs

J. BRONSON BAYLISS

A rectangular weir is a simple and inexpensive flow measuring device. Although it is limited to flow in open channels, it is well suited to measurement of waste streams or to any type of surface stream.

As it is economical to construct and install, it is often used for temporary measurements.

Flow of water through a sharp-edged, rectangular weir can be described by the Francis formula[1]:

$$q = 3.33 \, (L - 0.2H) \, H^{1.5}$$

where q = rate of flow, ft^3/sec

      L = length of weir crest (horizontal width of rectangular notch), ft

      H = head of water measured from weir crest, ft

Although the weir itself is simple, this empirical equation is rather unwieldly. It is well adapted to convenient calculation by nomograph. Figure 1-24a yields flow in gal/min when one aligns weir size in inches and head of water, also in inches.

The V-notch or triangular weir is slightly more complicated than the rectangular weir, but it provides considerably more flexibility. In general, the rectangular weir is suitable for measurement of large flows, but the V-notch is preferable for small flows or where a large range in head is desired.

Flow of water through a sharp-edged, V-notch weir can be described by the empirical equation[1]

$$q = 2.505 \, (\tan \alpha/2)^{0.996} H^{2.47}$$

where q = rate of flow, ft^3/sec

      α = vertical angle of notch, degrees

      H = head of water above vertex of notch, ft

Figure 1-24b has a folded scale for head and flow to provide range of flow from 3 to 3000 gal/min. Rate of flow on high or low range scale is read by aligning the proper V-notch angle with corresponding range on the head scale.

---

[1]Perry, J. H., *Chemical Engineers' Handbook*, p. 408, 3rd ed., McGraw-Hill Book Co., Inc., New York, 1950.

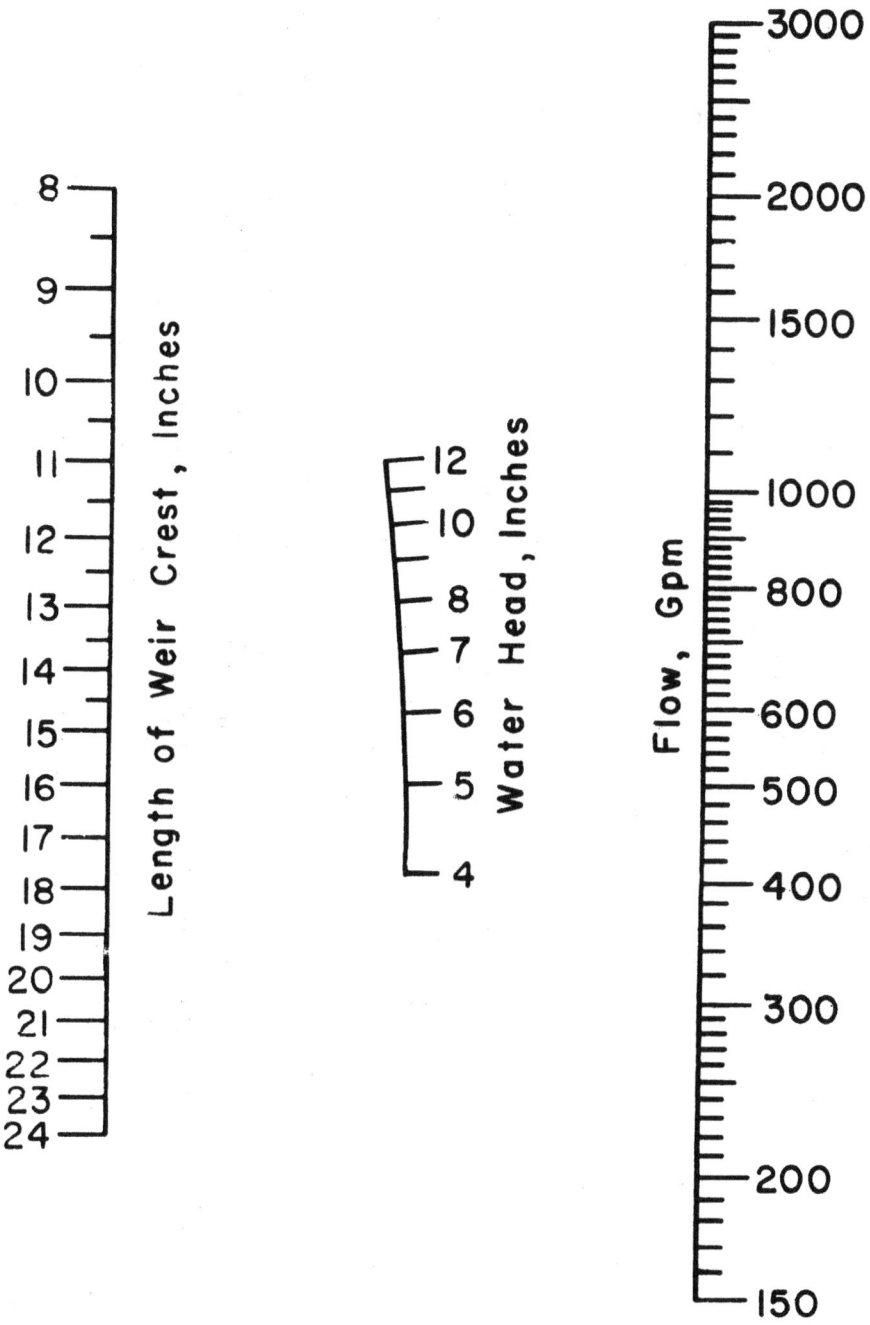

Figure 1-24a

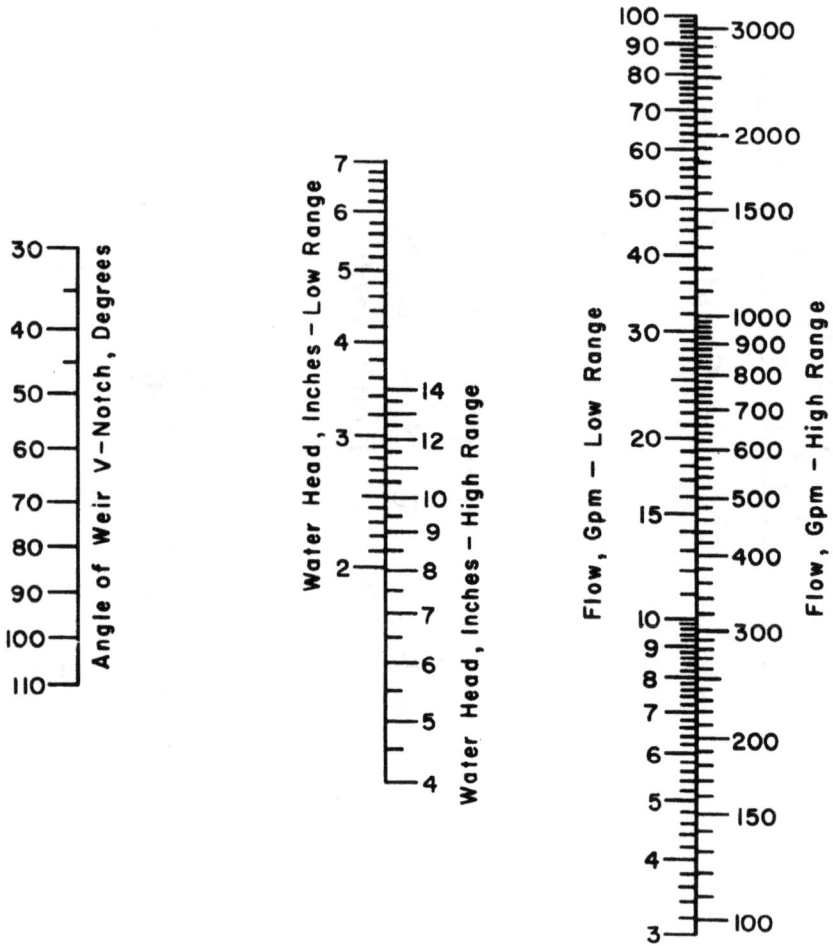

Figure 1-24b

UNIT 2

Some Aspects of Heat

Heat Transfer Coefficients—Specific Heats—Industrial Requirements

Most chemical processes are concerned with some aspects of heat. In this unit, nomographs that deal with heat transfer by conduction, convection, and radiation are supplemented by others that cover nucleate boiling, specific heats, heat content, and sensible heat loss.

2-1 Heat Transfer Coefficients for Nucleate Boiling

IRVING GRANET

Nucleate boiling has become one of the most frequently used methods of heat transfer in the chemical and allied fields. In particular most nuclear reactors depend upon this phenomenon to generate steam.

Glasstone[1] gives a formulation for the heat transfer to water in nucleate boiling as a function of pressure and temperature differential. In reduced form, this equation is

$$h = C (\Delta t)^{1.42}$$

where h = coefficient of heat transfer, Btu/(hr)(ft²)(°F)

       c = a constant dependent upon pressure

     Δt = temperature difference between surface causing boiling and saturation temperature of fluid being boiled, °F

Figure 2-1 is a solution of this equation for various pressures.

Typical Example

Heat is transferred to water at 215 lb/in.² abs. from a tube. The surface of the tube is 10°F hotter than the saturation temperature of the water. Determine the heat transfer coefficient.

Connect 10° on the Δt-scale with 215 lb/in.² abs. on the pressure scale and read h as 2170 Btu/(hr)(ft²)(°F).

[1]Glasstone, S. *Principles of Nuclear Reactor Engineering*, 1st ed, pp. 694 *et seq.*, D. Van Nostrand Company, Inc., Princeton, N.J., 1955.

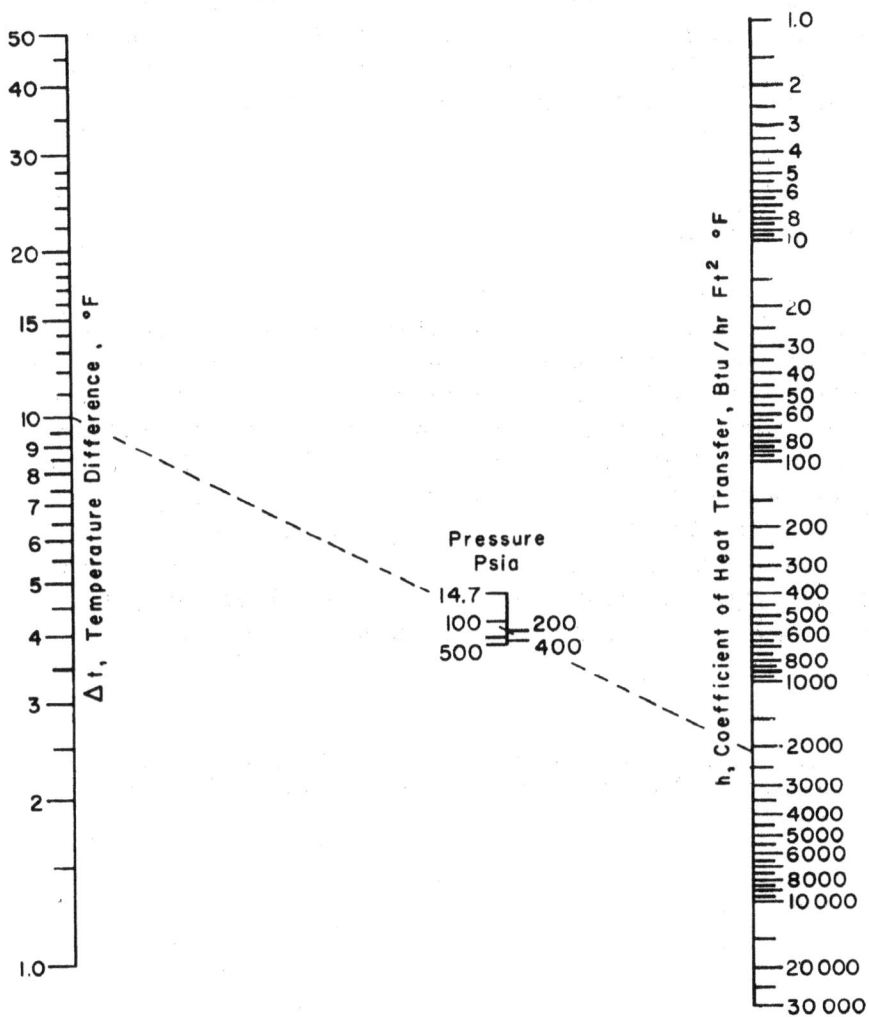

Figure 2-1

2-2   Heat Transfer by Convection From
Various Surfaces

ELIZABETH SHROFF

Figure 2-2 offers a convenient means of solving for the coefficient of heat transfer by natural convection (conduction and convection) from outer surfaces of horizontal and vertical pipes and plates to air at atmospheric pressure and ordinary temperatures.

Typical Example

For solving Equations 1, 2, and 5[1] on chart, connect the value of Δt, temperature potential in °F, with a point on the d_0-scale (diameter, in in.) to intersect the reference line. From this point draw a line to the corresponding circled reference point for the equation and read the desired value of the surface coefficient of heat transfer, in Btu/(hr)(ft²)(°F) at the intersection with the h_c-scale.

For solving Equations 3, 4, and 6, draw a line from Δt to d_0 or to Z, the height of the plate in feet, to intersect the reference line. Connect this point with the corresponding circled reference point for the equation and read the desired coefficient at the intersection with the h_c-scale.

[1]Equations 1, 2, 3, and 6 appear on p. 474 of Perry's *Chemical Engineers' Handbook*, 3rd ed., McGraw-Hill Book Co., New York, N.Y., 1950.

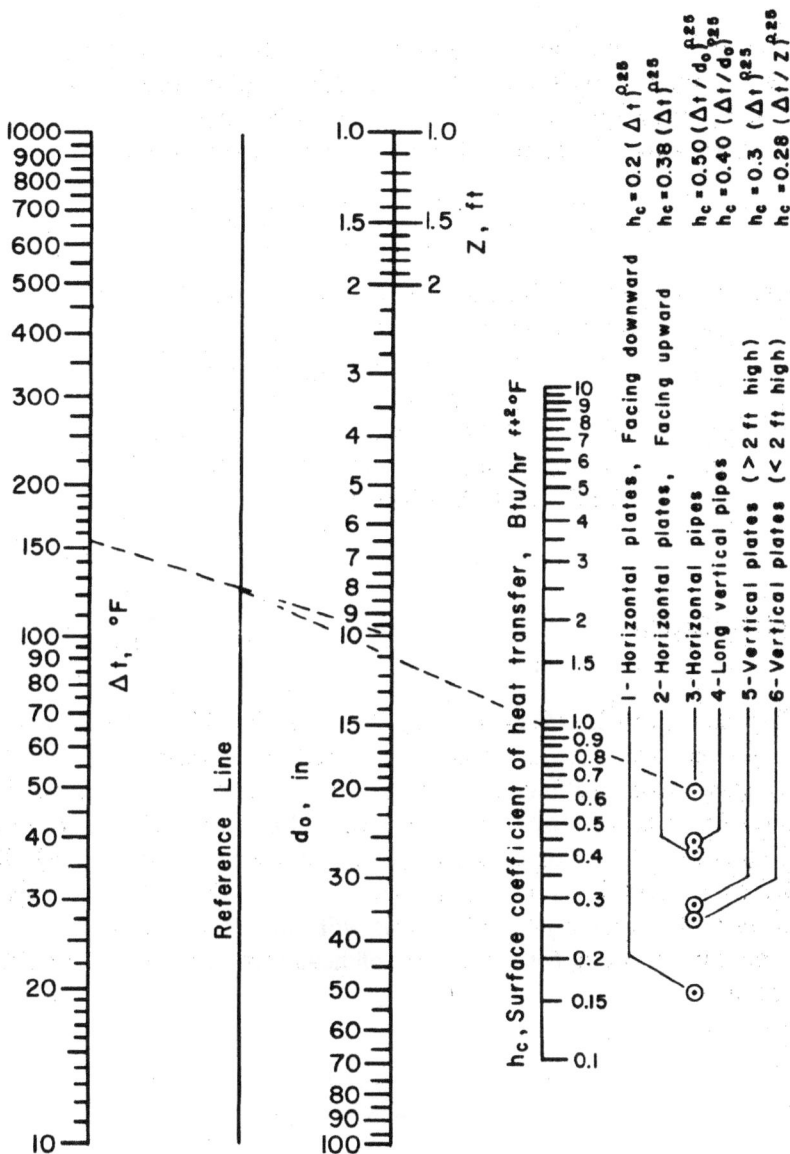

Figure 2-2

2-3   Sieder-Tate Film Coefficient for Liquids

GEORGE M. MACHWART

Film coefficients are used in calculations of heating and cooling of fluids in pipes and tubes. They are of various forms. One that is usually encountered is that of Sieder-Tate. This may be applied to a wide range of liquids. The usual form is as follows:

$$\frac{hD}{k} = 0.027 \left[\frac{DG}{\mu}\right]^{0.8} \left[\frac{C\mu}{k}\right]^{\frac{1}{3}} \left[\frac{\mu}{\mu_w}\right]^{0.14}$$

where h = film coefficient of heat transfer, $Btu/(hr)(ft^2)(°F)$
        D = inner diameter of pipe, ft
        k = thermal conductivity, $Btu(ft)/(hr)(ft^2)(°F)$
        G = mass velocity, $lb/(hr)(ft^2)$
        μ = viscosity of fluid, $lb/(ft)(hr)$
        μ_w = viscosity of fluid at pipe-wall temperature, $lb/(ft)(hr)$
        C = specific heat, $Btu/(lb)(°F)$

Typical Example

To use Figure 2-3, draw a straight line from 0.17 on the k-scale to 0.1 on the D-scale to intersect R_1. From intersection on the R_1-scale, draw a straight line to 1×10^3 on the G- scale to intersect the R_2-scale. From this intersection draw line to 1.0 on the C-scale to cross the R_3-scale. From this R_3-scale intersection pass line through R_4-scale to 10 on the μ-scale. Finally, draw line from R_4-scale intersection to 100 on the μ_w-scale. This will intersect the h-scale at 205, or the correct value for coefficient of heat transfer for the condition mentioned.

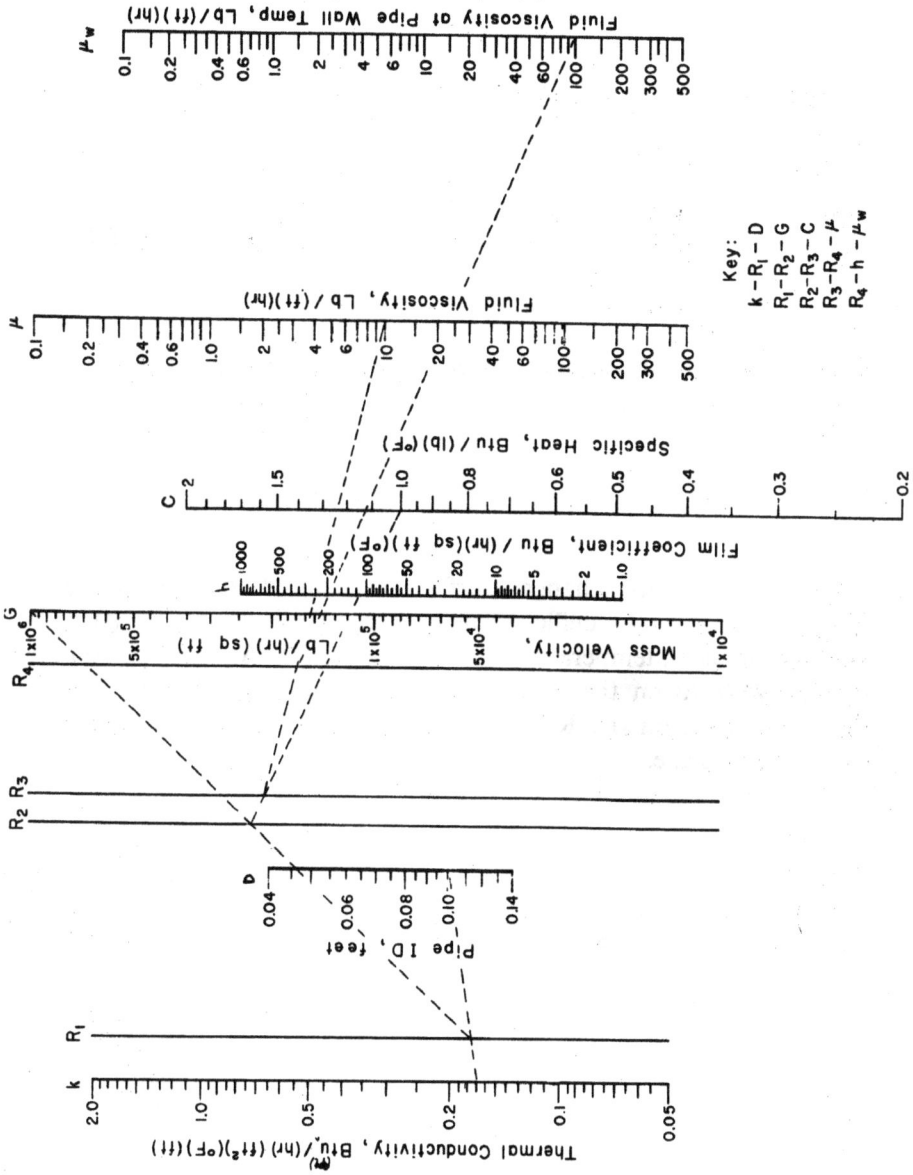

Figure 2-3

2-4   Heat-Exchanger Surface

F. CAPLAN

Figure 2-4 solves the equation

$$Q = UA\Delta t$$

where Q = heat transmitted, Btu/hr
       U = over-all coefficient of heat transfer, Btu/(hr)(ft²)(°F)
       A = heat-transfer surface, ft²
       Δt = temperature difference, °F

Note: For heat exchangers, use the logarithmic mean temperature difference.

Typical Example

How much heat-exchanger surface is required to remove 15,000 Btu/hr, if the overall coefficient of heat transfer is 80 Btu/(hr)(ft²)(°F) and the log-mean temperature difference is 70°F? Align 15,000 on the Q-scale with 70 on the Δt-scale and continue to the R-line. Align the intersection on the R-line with 80 on the U-scale and read 2.68 ft² on the A-scale.

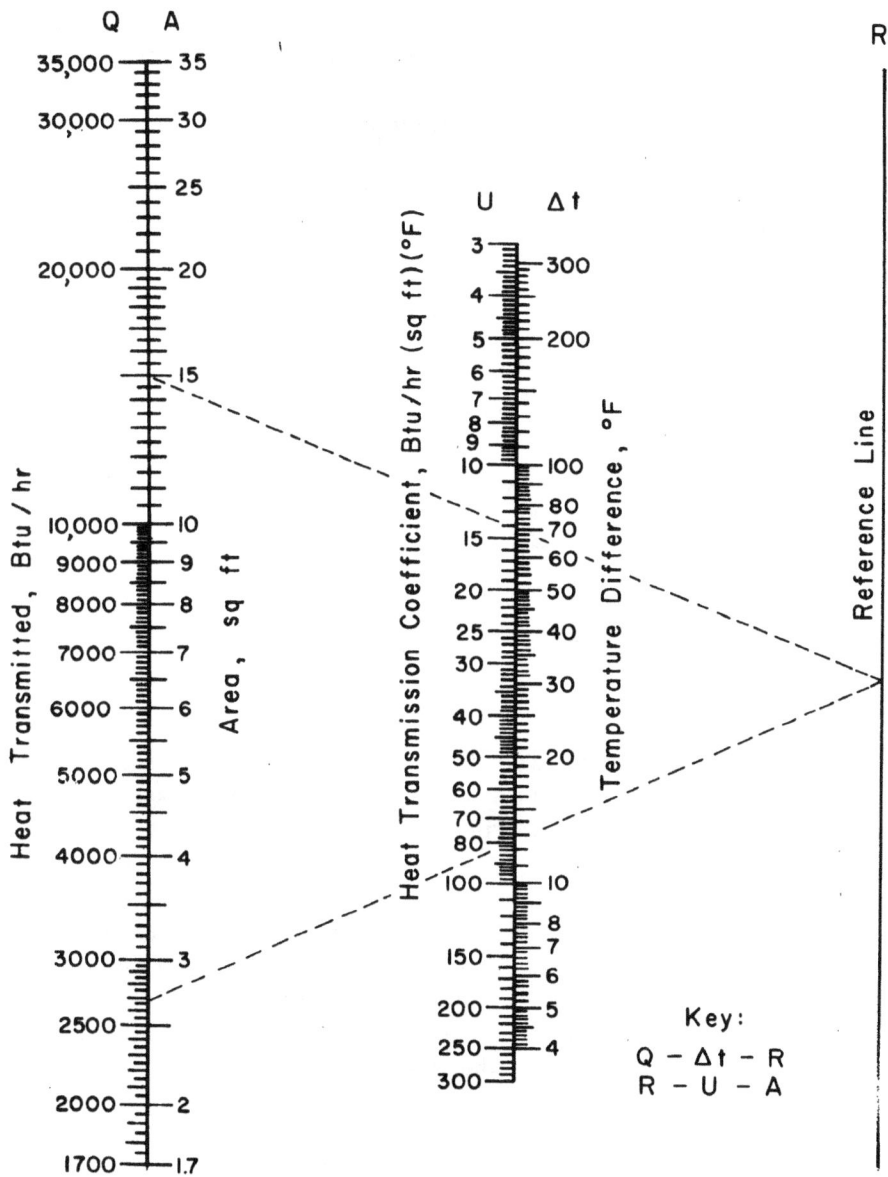

Figure 2-4

2-5   Heat Transfer Through Liquid Metals

P. D. SHROFF

Liquid metals are used extensively in industry as heating and cooling-media because of their high heat capacities, high thermal conductivities and wide useful-temperature ranges. Applications include heating baths, cooling of exothermic reactions and circulation through high-temperature kettles and nuclear reactors.

Figure 2-5 permits determination of heat-transfer coefficients for liquid metals in turbulent flow. It is based on the equation[1]

$$h = 73.43 \left[\frac{k}{D} \right]^{0.6} (u \, \rho \, c)^{0.4}$$

where   h   = heat-transfer coefficient, Btu/(hr)(ft²)(°F)
         k   = thermal conductivity, Btu/(hr)(ft)(°F)
         ρ = density, lb/ft³
         c   = heat capacity, Btu/(lb)(°F)
         u   = linear velocity
         D   = inside diameter of tube, in.

Typical Example

What is the heat-transfer coefficient for liquid tin flowing at a linear velocity of 10 ft/sec through a 1″-ID pipe when the average temperature is 600°F? Fluid properties for the metal at that temperature are k = 18.61 Btu/(hr)(ft)(°F), ρ = 418.7 lb/ft³, and c = 0.065 Btu/(lb)(°F).

Connect 0.065 on the c-scale with 418.7 on the ρ-scale, and note the intersection on the P-axis. Connect this point with 18.61 on the k-scale, and mark the intersection on the Q-axis. Now draw a line between this intersection point and 10 on the u-scale. Mark off the intersection of this line with the R-axis. Finally, connect this point on the R-axis with 1 on the D-scale, and read 4000 Btu/(hr)(ft²)(°F) on the h-scale.

[1]*Review of Experimental Investigations of Liquid-Metal Heat Transfer*, Lubarsky, B., and Kaufman, S. J., NACA TN 3336, National Advisory Committee for Aeronautics, March 1955.

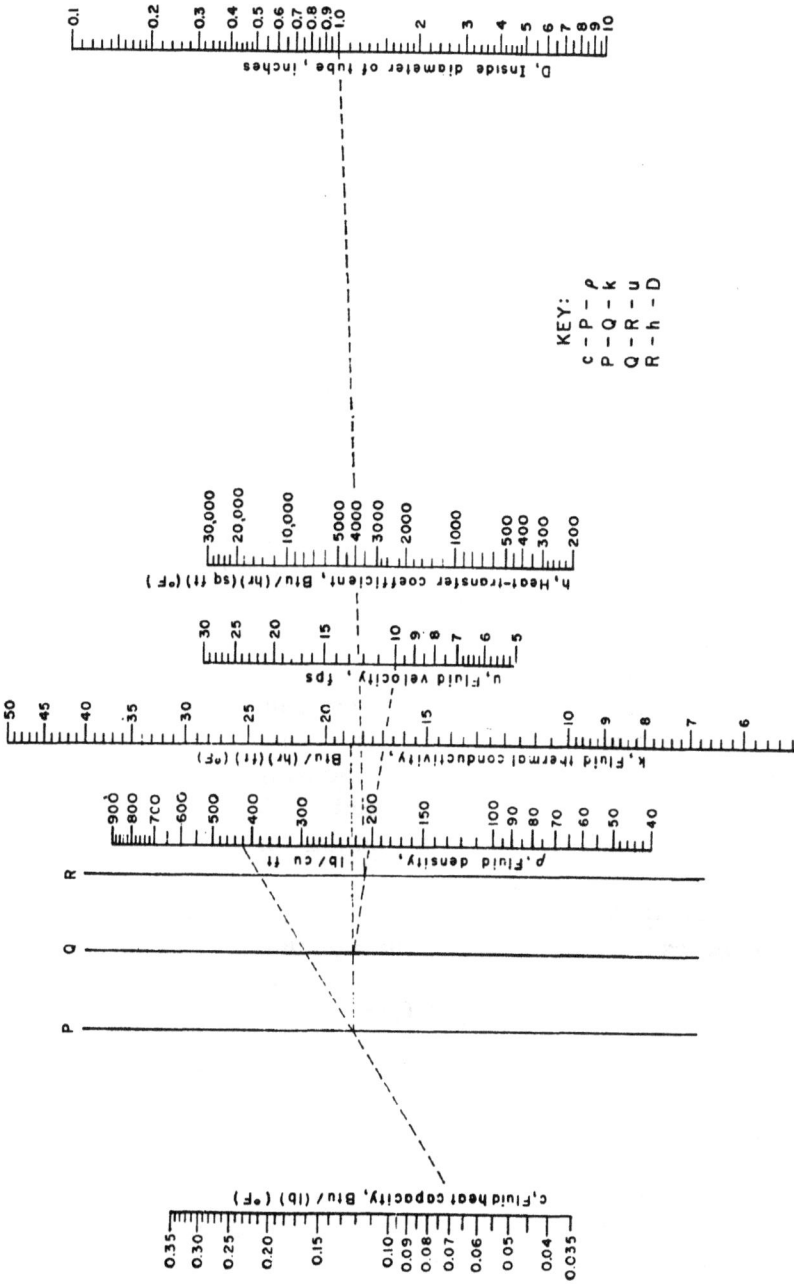

KEY:
c – P – ρ
P – Q – k
Q – R – u
R – h – D

D, Inside diameter of tube, inches

h, Heat-transfer coefficient, Btu/(hr)(sq ft)(°F)

u, Fluid velocity, fps

k, Fluid thermal conductivity, Btu/(hr)(ft)(°F)

ρ, Fluid density, lb/cu ft

c, Fluid heat capacity, Btu/(lb)(°F)

Figure 2-5

2-6   Radiant Heat Transfer in Cryogenic Processes

ELIZABETH D. SHROFF

In choosing proper cryogenic-container systems which use high-vacuum insulation, it is necessary to know the net rate of radiant heat transfer which will take place in the system. When vacuum insulation is used, the heat transferred from a surface at room temperature (300°K) to one at 77°K (boiling point of nitrogen) or lower occurs almost entirely by radiation. This condition is a factor when considering ordinary container systems for liquid oxygen or nitrogen, or for unshielded container systems to accommodate other very-low-boiling-point liquefied gases.

In all these typical cases, the net rate of radiant heat transfer is unaffected for practical purposes when the temperature of the cold surface is reduced below 77°K (down to such temperatures as the boiling point of hydrogen, 20°K, or that of helium, 4.2°K). This is because the temperature of the warm surface is so much greater than 77°K at the start. Figure 2-6a is to be used when the higher temperature is 300°K.

Another situation frequently encountered in low-temperature work involves radiant heat transfer from a surface at 77°K to one at 20°K or lower. These conditions are encountered in vessels holding liquid hydrogen or helium which have protective shields cooled with liquid nitrogen. Figure 2-6b gives the solution for the net radiant heat transfer under these conditions.

The nomographs are based on the general correlation for net rate of radiant heat transfer given by the equation

$$\frac{Q}{A_1} = \frac{\sigma\,(T_w{}^4 - T_c{}^4)}{\dfrac{1}{\varepsilon_1} + \dfrac{A_1}{A_2}\left[\dfrac{1}{\varepsilon_2} - 1\right]}$$

where Q/A_1 =   rate of heat transferred, watts/ft^2
        T_w =   temperature of warm surface, °K
        T_c =   temperature of cold surface, °K
        σ =   the Stefan-Boltzmann constant, 0.533×10^{-8} (watts/(ft^2)(°K^4))
  $\varepsilon_1 = \varepsilon_2$ =   surface emissivities
        A_1 =   inner area, ft^2
        A_2 =   outer area, ft^2

Typical Examples

1) What is the net rate of radiant heat transfer when the outer surface is at room temperature, the inner surface is at the temperature of liquid nitrogen, the area ratio is 0.8 and the surface emissivities are assumed to be 0.033?

Use Figure 2-6a and join 0.8 on the A-scale with 0.033 on the E-scale to give on the ψ-scale a net rate of radiant heat transfer of 0.80 watts/ft² of inner surface area.

2) What is the net rate of radiant heat transfer if the outer surface is at 77°K, the inner surface is at 20°K or lower, the area ratio is 0.92 and the emissivities are assumed to be 0.06? Figure 2-6b and the same process as in the previous example (noting that scales with the same subscripts must be used together), the solution is 0.006 watts/ft² of inner surface area.

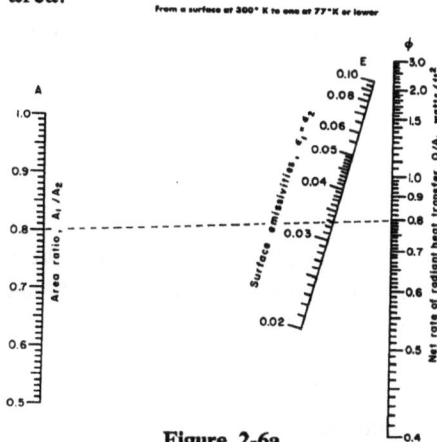

Figure 2-6a

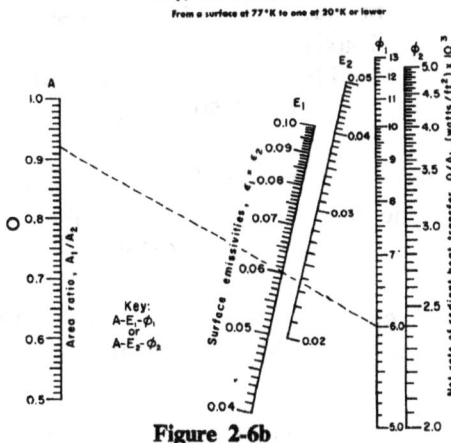

Figure 2-6b

2-7   Heat Losses from Horizontal Pipes

IRVING GRANET and D. S. DAVIS

Heat losses by natural convection and radiation from bare or insulated horizontal steel pipe to ambient air at or near 80°F can be calculated by means of Figure 2-7. In agreement with data[1] in the literature, the chart is based on the specially developed equation

$$L = (10^{a-b \Delta t}) \pi (D+2x) \frac{P \Delta t}{12}$$

where L = heat loss, Btu/hr

Δt = temp of outer surface of pipe or insulation minus temp of surrounding air, °F

D = outer diameter of bare pipe, inches

x = thickness of insulation, inches

P = length of pipe, ft

a and b depend upon outer diameters of Schedule 40 pipe

Typical Example

Use of Figure 2-7 is illustrated as follows: At what rate is heat lost to ambient air at 80°F from 50 ft of 4 in. Schedule 40 steel pipe that has 1½ in. of insulation if the temperature of the outer surface of the insulation is 270°F?

Note that D + 2x = 4.50 + 2(1.50) = 7.50 inches.

Following the key, connect 270°−80° or 190° on the Δt_1-scale and 4 on the nominal pipe size scale with a straight line and mark the intersection with the L-axis.

Follow the slanted guide lines for a first intersection with the β-axis and connect this point with 7.50 on the (D+2x)-scale, noting the intersection with the α-axis.

Connect this point with 50 on the P-scale and mark the second intersection with the β-axis. Connect this point with 190 on the Δt_2-scale and read the rate of loss of heat as 5×10^4 or 50,000 Btu/hr on the L-scale.

For a value of P of 500 ft instead of 50 ft, multiply the value of L by 10 to obtain 5×10^5 or 500,000 Btu/hr.

[1]Marks, L. S., ed., *Mechanical Engineers' Handbook*, 5th ed., p. 377, McGraw-Hill Book Co., Inc., New York, 1951.

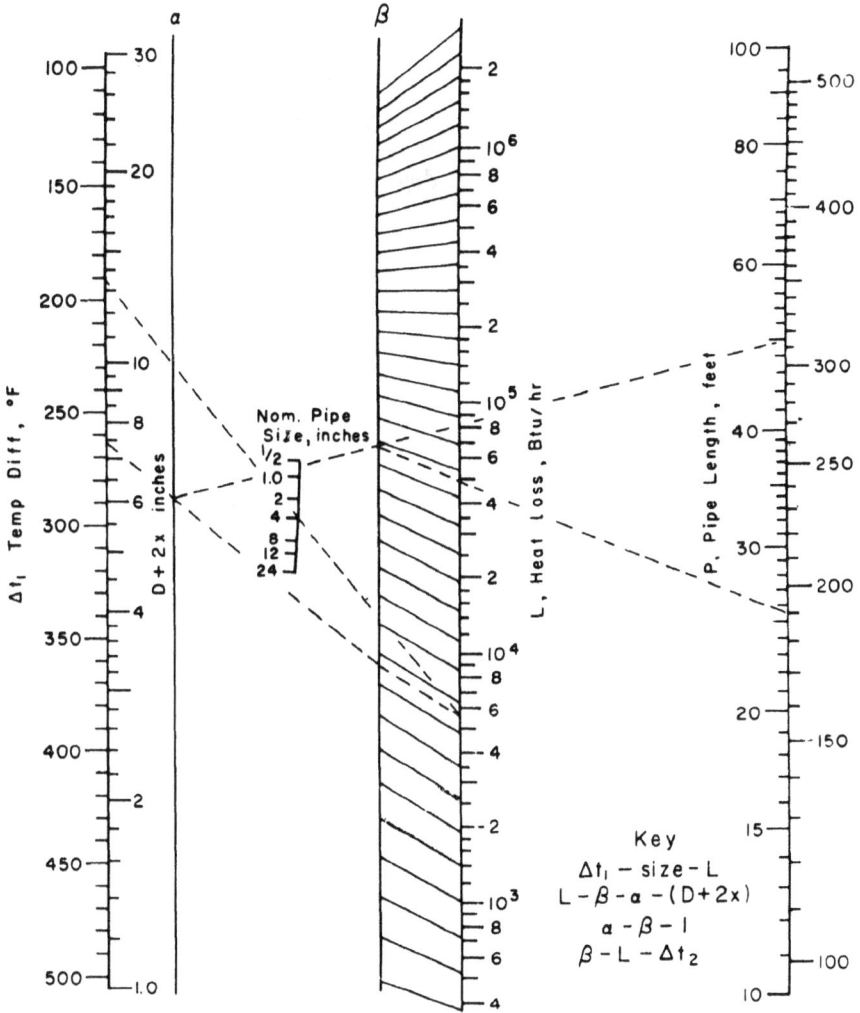

Figure 2-7

2-8   Total Heat Content of Water Vapor and Carbon Dioxide

GEORGE E. MAPSTONE

For flame-temperature calculations, it is necessary to know the total heat content of the various products of combustion above 60°F. Figure 2-8, based on sound data,[1] enables the total heat content of water vapor and of carbon dioxide at 1 atm to be readily obtained, knowing the temperature and the degree of dissociation of these compounds.

Typical Example

What is the total heat content above 60°F of carbon dioxide gas at 3650°F if it is 11% dissociated? Connect 11% on the dissociation (left-hand) scale with 3650°F on the carbon dioxide (left-center) scale, and extend this line to obtain 158 Btu/ft³ on the heat-content (right-hand) scale. Proceed similarly for water vapor.

[1]*Fuel Flue Gases*, pp. 58-59, American Gas Association, New York, 1941.

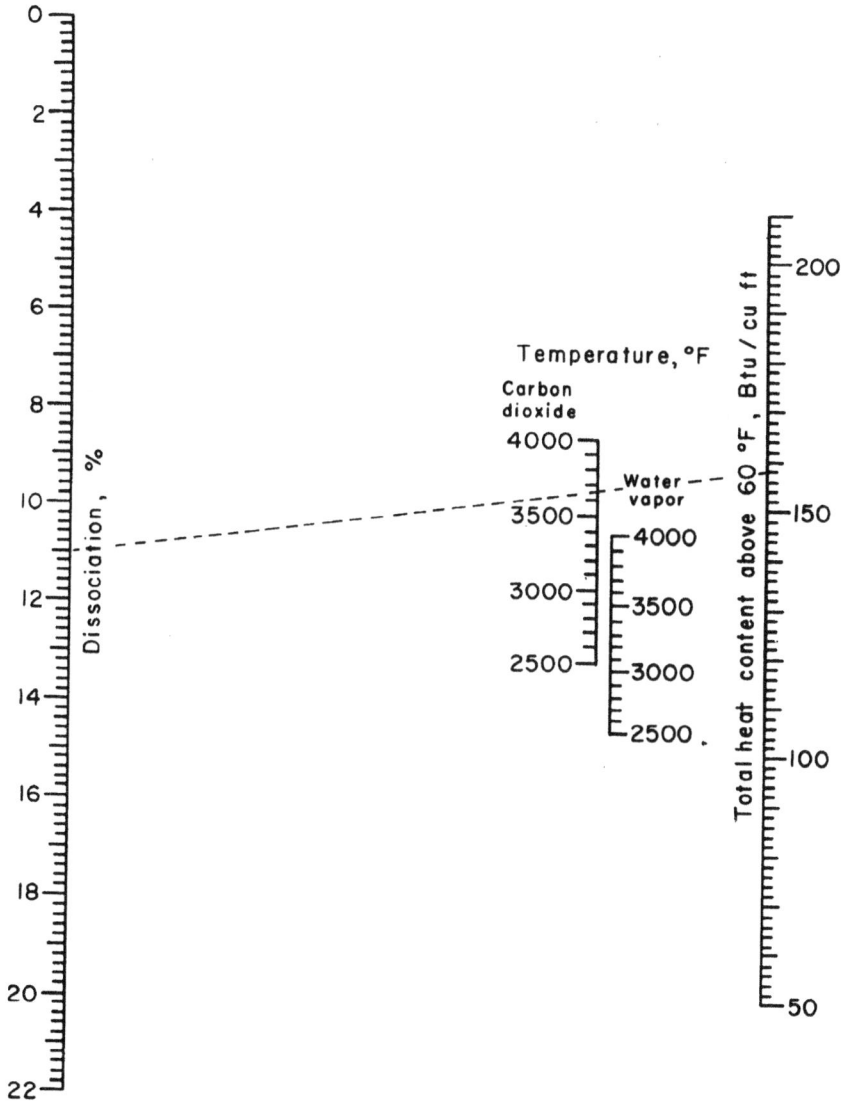

Figure 2-8

2-9   Specific Heats of Metals and Quartz

JOHN B. GARDNER

To find the specific heat of a particular solid with Figure 2-9, draw a straight line through the reference point from the desired temperature. The intercept on left-hand scale is specific heat in g-cal/(g)(°C) or Btu/(lb)(°F).

For example, to find the specific heat of iron at 600°C, draw line from 600°C through the reference point (10). The intercept gives the value of 0.1735 g-cal/(g)(°C) on the specific heat scale.

Chart Number	Solid	Usable Temperature, °C
1	gold	0—1000
2	lead	0— 300
3	silver	0— 900
4	copper	0—1100
5	zinc	0— 400
6	zinc	500—1000
7	nickel	0— 300
8	nickel	400—1400
9	cobalt	0—1400
10	iron	0— 700
11	aluminum	0— 600
12	quartz	600—1400

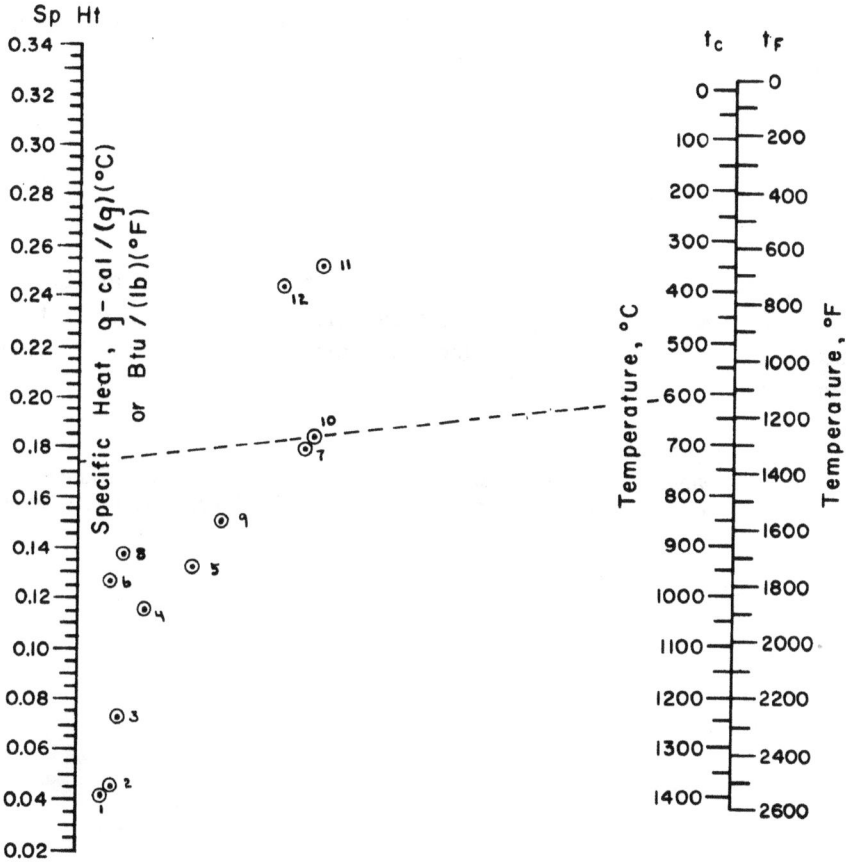

Figure 2-9

2-10   **Specific Heats of Aqueous Nitric**
Acid Solutions

ABE DEVORE

The specific heats of aqueous nitric acid solutions can be rapidly determined by use of Figure 2-10. Within the temperature limits represented on the chart, the specific heat varies linearly with temperature in accordance with sound data[1].

Typical Example

What is the specific heat of a 50% nitric acid solution at 100°F? Connect·100 on the temperature scale and 50 on the concentration scale with a straight line. Extend the line to the specific-heat scale and read the value as 0.697 Btu/(lb)(°F).

[1]Timmermans, Jean, *The Physicochemical Constants of Binary Systems in Concentrated Solutions*, Vol. 4, p. 517, Interscience Publishers, Inc. N.Y. (1960).

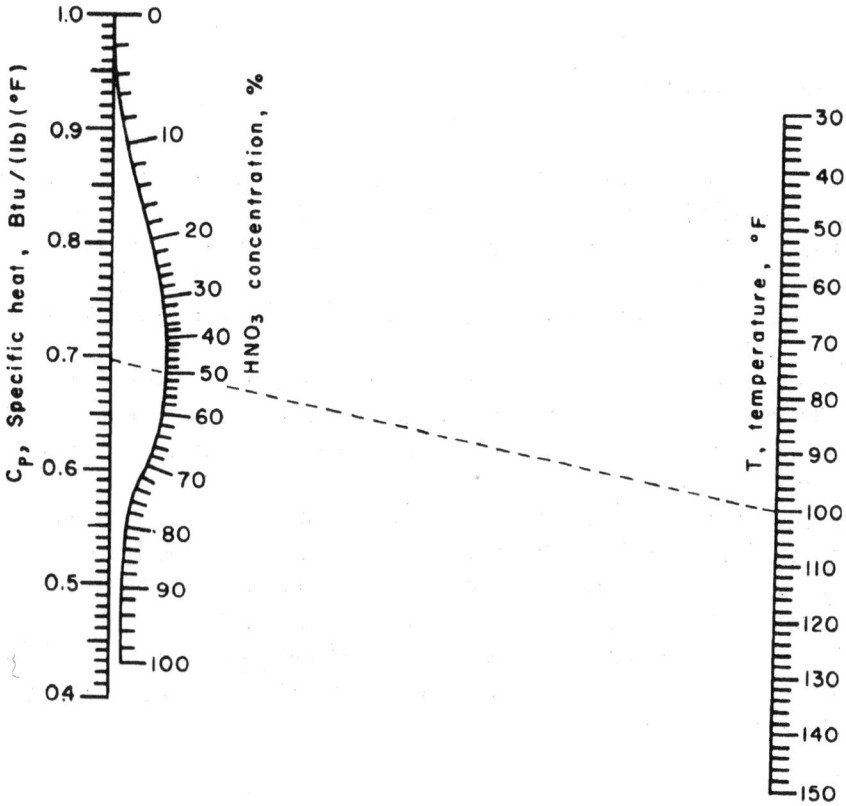

Figure 2-10

2-11   Evaluating Resistance Heaters

HARRY D. WILLS SR.

Figure 2-11 permits rapid evaluations of heaters and transformers in industrial electric-heating applications.

Typical Example

Determine the compatibility of a 5-kw, 220 V heater and a 28-amp, 270 V variable autotransformer. Also, find the maximum power, and determine which of the two units is the more susceptible to failure.

Draw a straight line from 5 on the W-scale to 220 on the E-scale. Intersection of this line with the I_1-scale yields a heater-current value of 23 amp.

Next connect 220 on the E-scale with 23 on the I_2-scale and find the heater resistance to be 9.7 Ω. To find the maximum current, extend a line from 270 on the E-scale through 9.7 on the R-scale to obtain I_2max = 28 amp.

Now find the transformer rating on scale W to be 7.6 kw by extending a line connecting 270 on the E-scale and 28 on the I_1-scale.

The transformer is compatible with the heater, for at maximum voltage (270 V), the current will not exceed the transformer rating (28 amp). The heater is most likely to fail at maximum power if the heat is not removed at a sufficient rate to keep its temperature at a safe level.

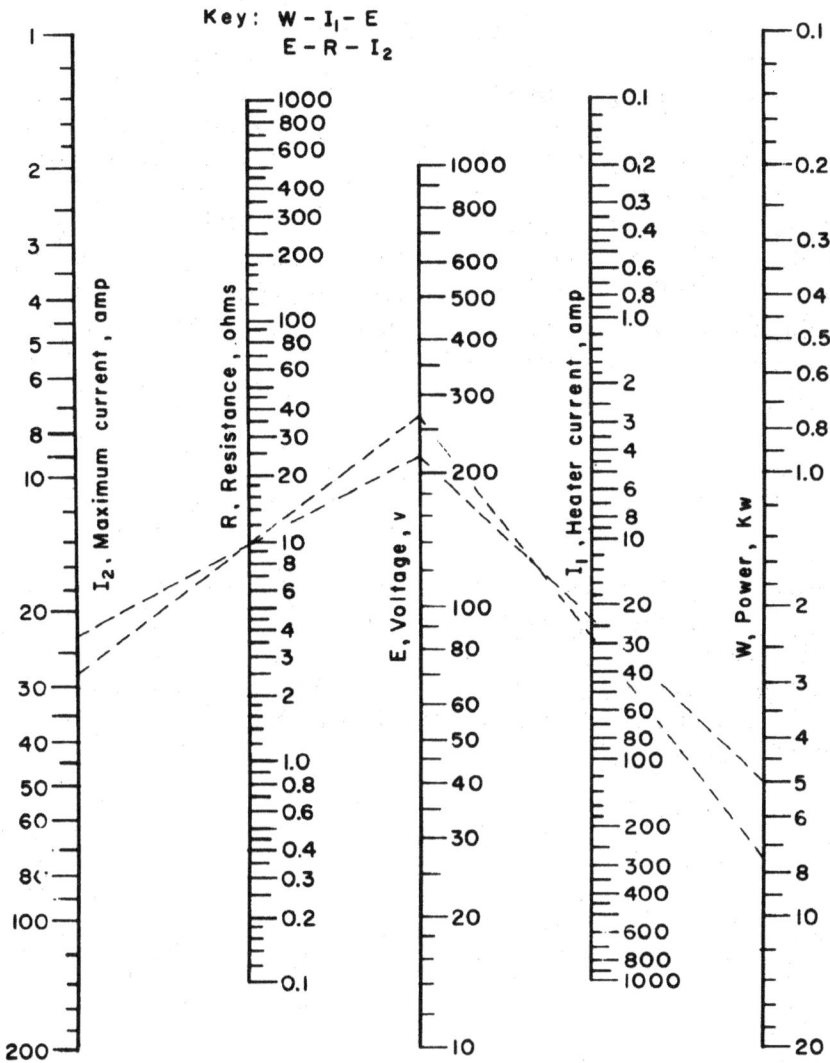

Key: W – I₁ – E
E – R – I₂

Figure 2-11

2-12   Estimating Heat Loss in Boiler Flue Gases (Bituminous Coal as Fuel)

GEORGE E. MAPSTONE

As the principal purpose of operating a boiler is to transfer heat to water or some other medium, heat used to increase temperature of flue gases is wasted. Air in excess of the amount needed for complete combustion increases such losses. If the fuel composition is known, the amount of excess air can be calculated through flue gas analysis. Flue gas temperature is easily measured. Figure 2-12 shows both the excess air and the heat loss, when temperature and analysis of flue gas are known. It is designed specifically for bituminous coal with a heat content of 14,400 Btu/lb.

Typical Example

Flue gas contains 10.1% carbon dioxide and has a temperature of 550°F. How much excess air is being used and what is the heat loss to flue gases?

Using a straightedge, connect 550°F on the temperature-scale with 10.1% CO_2. Intersection of the straightedge shows 20% loss in fuel and 80% excess of air.

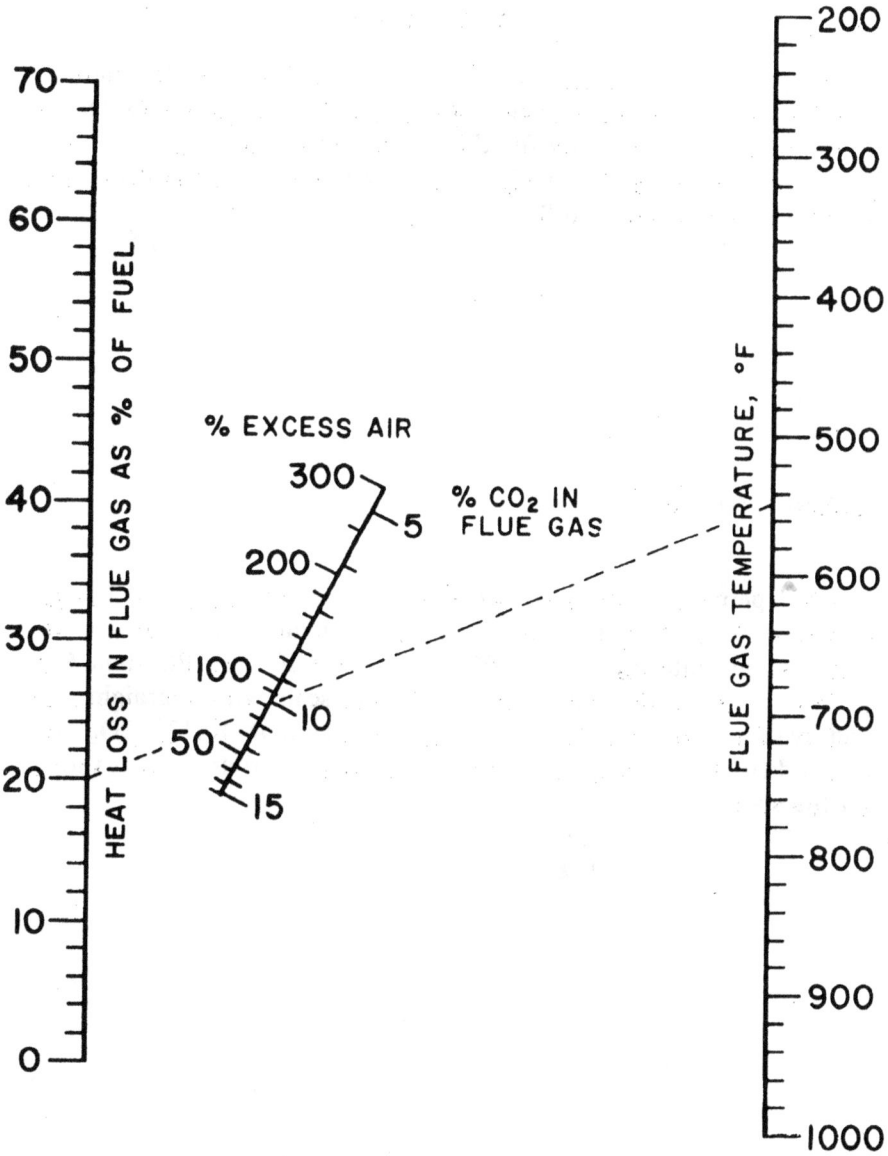

Figure 2-12

2-13   Sensible Heat (Dry-Gas) Loss from Oil-Burning Boilers

VICTOR J. COTZ

Dry-gas (sensible heat) loss is the major loss in a boiler, and the one best controlled by proper boiler operation. Figure 2-13 permits estimation of this quantity for the combustion in a boiler furnace of No. 6 fuel oil, which has a higher heating value of 18,400 Btu/lb and an ultimate analysis as follows:

C	87.8 %
H_2	10.33
O_2	0.5
N_2	0.14
S	1.16

Typical Example

What percentage dry-gas loss can be expected when the flue gas contains 10.7% CO_2 and the difference in temperature between the flue gas and entering air is 300°F? Connect 10.7 on the slant CO_2 scale and 300 on the temperature-difference scale with a straight line, and read the dry-gas loss as 7.5%. (Note that 10.7% CO_2 corresponds to 40% excess air supplied to the boiler furnace unit as shown on the slant axis.)

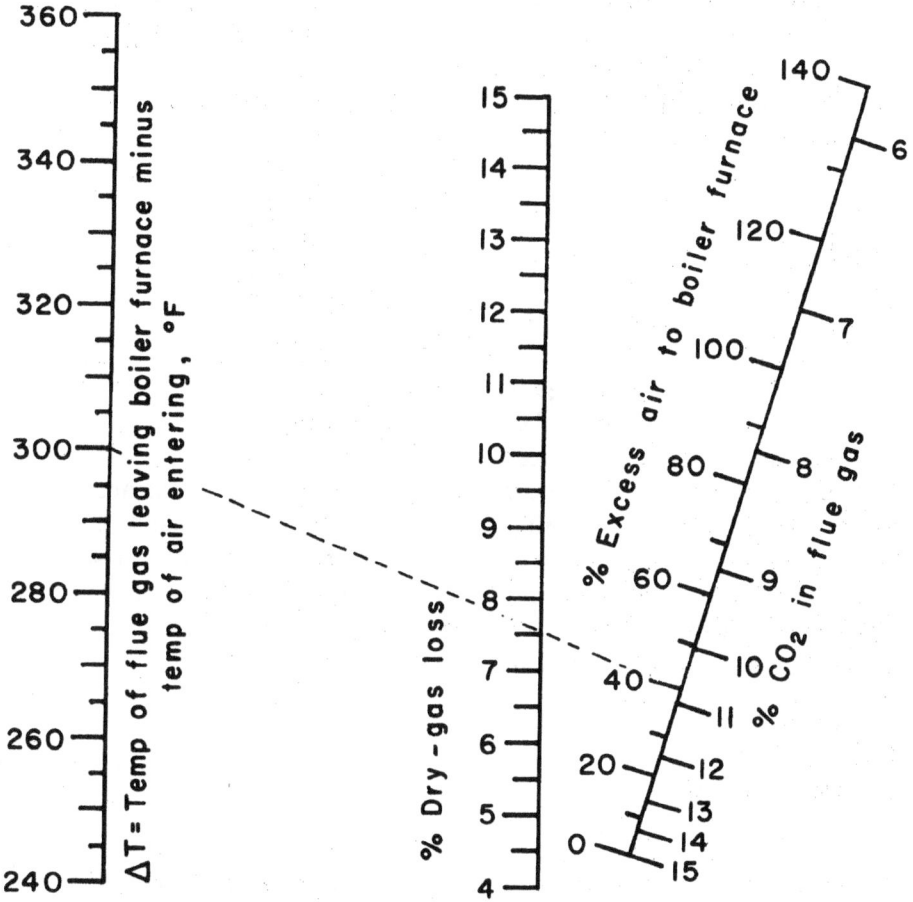

Figure 2-13

2-14   Steam-Trap Warming-Up Load

WILLIAM RESNICK

One design criterion for steam traps on steam mains and risers is that they be sized to handle condensate formed when steam is turned on to the main. Calculation of amount of steam condensate formed, in raising pipe from room temperature to saturated steam temperature, is simplified by use of Figure 2-14.

This nomograph is based on the heat-balance equation

$$S = \frac{C_p P(T_r - T_s)}{\lambda}$$

where S = steam condensate, lb/lineal ft of pipe
C_p = average heat capacity of steel, 0.12 Btu/(lb)(°F)
P = weight of pipe, lb/lineal ft
T_r = room temperature, taken as 75°F
T_s = saturated steam temperature, °F
λ = latent heat of steam at T_s, Btu/lb

Data for P and λ were taken from standard sources.[1,2]

To use Figure 2-14, a straight line is drawn from the steam-temperature scale to the nominal-pipe-size scale and extended to the right-hand scale. The reading on this scale will be amount of steam condensed per lineal ft of pipe, in raising pipe from 75°F to the saturated steam temperature. Radiation losses and warming-up load because of flanges and fittings are not included.

Typical Example

How much condensate must be removed from 100 ft of 10 in. Schedule 160 steam main when 450°F steam is turned on to main? From the nomograph, condensate load will be 6.6 lb/lineal ft of pipe. Warming-up load for main will be 100 × 6.6 or 660 lb.

[1]Perry, J. H., *Chemical Engineers' Handbook*, p. 416, Third Edition, McGraw-Hill Book Company, New York, 1950.
[2]Keenan, J. H., and F. G. Keys, *Thermodynamic Properties of Steam*, John Wiley & Sons, New York, 1936.

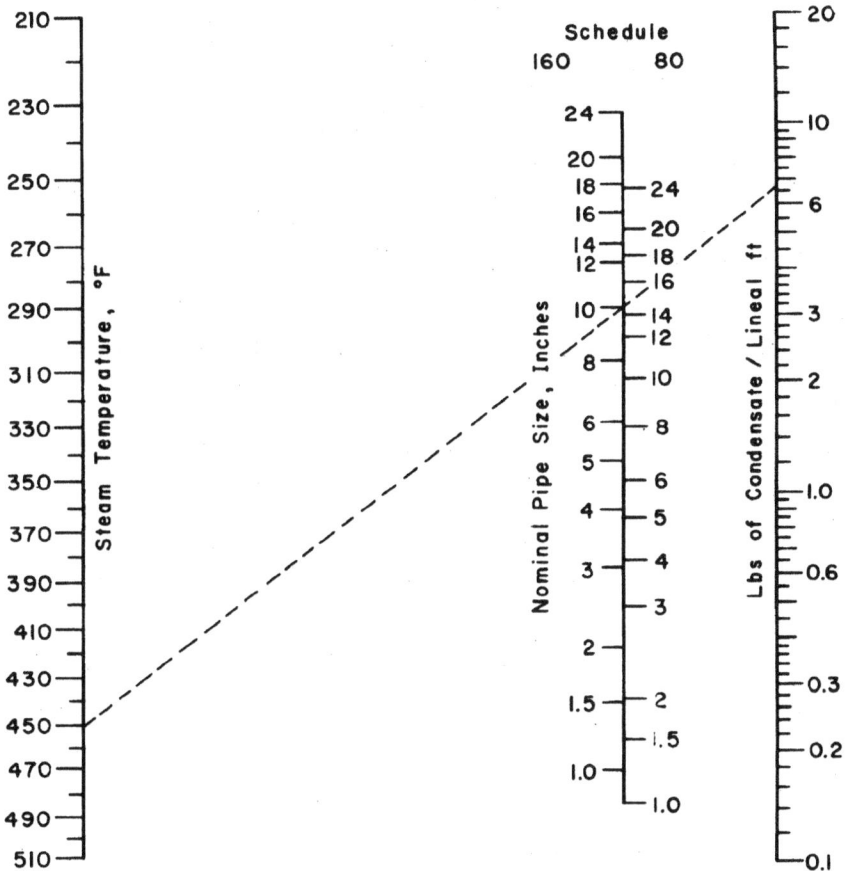

Figure 2-14

2-15   Heat Required for Melting Glass

D. S. DAVIS

Figure 2-15 permits rapid and accurate solution of the Hartford-Empire standard practice formula[1] for melting glass in regenerative furnaces:

$$H = 0.6\,A + 4.8\,G + 50$$

where   H = heat required, millions of Btu/day
      A = area of melting end of furnace, ft^2
      G = glass melted, tons/day

The broken line on the chart shows that 1250 million Btu are required to melt 170 tons of glass per day when the area of the melting end of the regenerative furnace is 640 ft^2.

[1]Creasy, M. S., and Lyle, *A. K.*, *Ceramic Industry*, **67** (3) 99, 1956.

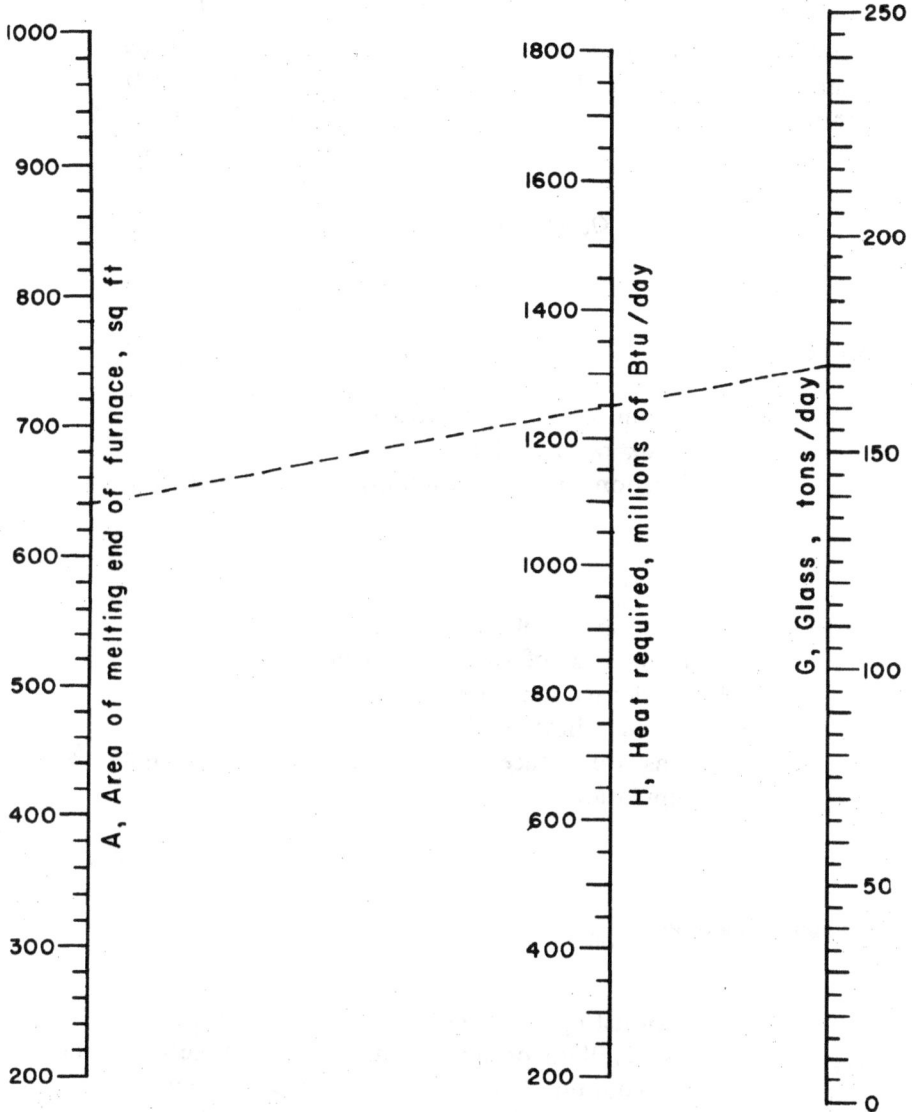

Figure 2-15

2-16   Estimation of Condenser and Reboiler
Heat Loads in Fractionators

P. D. SHROFF

For distillation systems where the simplifying McCabe-Theile assumptions can be made, the quantities of heat that must be transferred out of the solution (condenser heat load) and transferred into the solution (reboiler heat load) are given by the equations[1]

$$Q_c = F(R+1) \left[\frac{X_F - X_B}{X_D - X_B} \right] \Delta h_v$$

$$Q_R = F \left[\left[\frac{X_F - X_B}{X_D - X_B} \right] R + q \right] \Delta h_v$$

where X_F = feed composition, fraction %
X_D = distillate composition, fraction %
X_B = bottom composition, fraction %
R = reflux ratio (L/D)
q = function of feed enthalpy, defined by Treybal as

$$\frac{L' - L}{F}$$

F = feed rate, moles/hr
Δh_v = latent heat of vaporization of V', Btu/mole
Q_C = condenser heat load, Btu/hr
Q_R = reboiler heat load, Btu/hr

These equations are reduced to nomographic form (Figure 2-16) for ease of manipulation.

Typical Example

A feed stock consisting of 35 mole% A and 65 mole% B is to be separated into a distillate product containing 95 mole% A and a bottom product containing 5 mole % A. Estimate the condenser and reboiler heat loads for a feed rate of 100 moles/hr and a reflux ratio of 3.0. Use a value of 3000 Btu/mole for the latent heat of vaporization and 3.0 for q.

Condenser heat load. Join the intersection of X_D = 0.95 and X_B = 0.05 with X_F = 0.35 to intersect reference line A. From this point join 3.0 on the R-scale to intersect reference line C. Now connect this point with $\Delta h_V = 3000$ to intersect reference line E. Join 100 on the F-scale with the reference-line-E intersection point and read on the Q_C-scale the condenser heat load of 4.0×10^5 Btu/hr.

Reboiler heat load. Join the intersection of $X_D = 0.95$ and $X_B = 0.05$ with $X_F = 0.35$ to intersect reference line A. From this point join 3.0 on the R-scale to intersect reference line B. Connect this point with $q = 3.0$ to intersect reference line C. Now connect this point with $\Delta h_V = 3000$ to intersect reference line E. Join 100 on the F-scale with the reference-line-E intersection point and read on the Q_R-scale the rebolier heat load of 1.2×10^6 Btu/hr.

[1]J. L. Beckner, "One Step Equations to Find Re-boiler and Condenser Duty," *Chemical Engineering*, pp. 164-166, Feb. 20, 1961.

Note: For a feed rate F_1 greater than 100 moles/hr, the heat loads are obtained as before for 100 moles/hr and the results are mulpiplied by the ratio $F_1/100$. For example, in this illustration, for a feed rate of 160 moles/hr, $Q_c = (1.6)(4.0)(10^5) = 6.4 \times 10^5$ Btu/hr, and $Q_R = (1.6)(1.2)(10^6) = 1.92 \times 10^6$ Btu/hr.

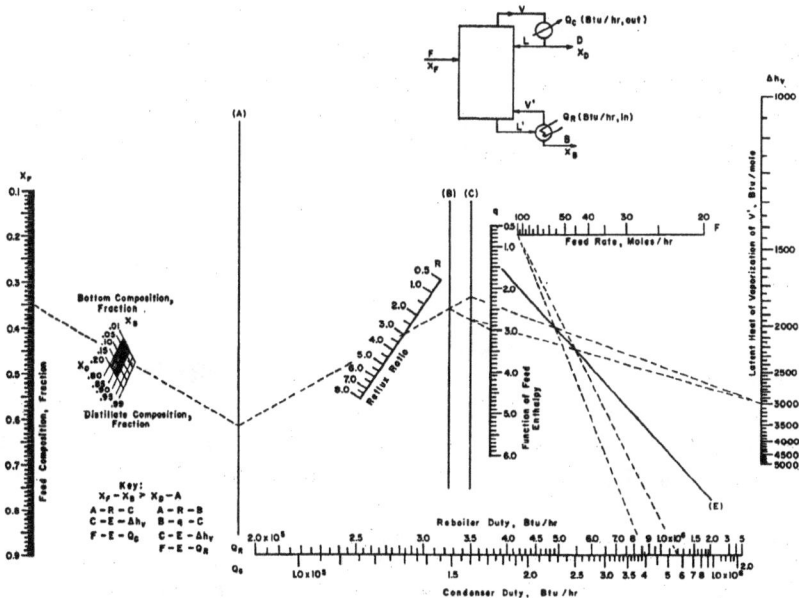

Figure 2-16

2-17   Effect of Velocity on Heat-Transfer Rates

ELIZABETH SHROFF

An increase in velocity of a stream in an exchanger causes heat-transfer rate to be improved, usually resulting in smaller units being required. This also results in increased pumping costs because of increase in pressure drop. It is therefore important to determine changes in heat-transfer rate and pressure drop arising from increase in velocity.

For heat transfer by convection involving liquid streams being heated or cooled where Re. No. exceeds 2100, and where no change of phase takes place, h_i, tube-side rate, will change as mass or linear velocity to 0.8 power; h_0, shell-side rate, will change as mass or linear velocity to 0.55 power.

It follows that

$$h_2 = h_1 (f_v)^n$$

where h_1, h_2 = coefficients of heat transfer, Btu/(hr)(ft^2)(°F) old and new, respectively

f_v = factor by which velocity is increased

n   = 0.80, for tube-side coefficient of heat transfer

n   = 0.55 for shell-side coefficient of heat transfer

From basic pressure drop equation, it is known that pressure drop will vary as square of velocity. These equations are represented in Figure 2-17 for ease of handling.

Typical Examples

Example 1:   Shell-side coefficient of heat transfer in an exchanger is 25 Btu/(hr)(ft^2)(°F). If velocity of liquid in shell is doubled, what is new shell rate? Connect 2.5 on h-scale with 2 on the f_v-scale. Read 3.66 on h_0-scale. New shell rate is 36.6 Btu/(hr)(ft^2)(°F) (line a).

Note: In above example, (a) if original shell rate is 250 Btu/(hr)(ft^2)(°F), new shell rate would be 366 Btu/(hr)(ft^2)(°F); (b) if original tube rate is 25 Btu/(hr)(ft^2)(°F) and velocity through tubes is doubled, new tube rate is found at intersection of line (a) with h_i-scale. It is equal to 43.5 Btu/(hr)(ft)(°F). Or, h_i of 250 would increase to 435; (c) since velocity is doubled, shell-side pressure drop would increase by factor of 4 as found on f_v-scale.

Example 2 — If shell-side velocity were reduced 50% ($f_v = \frac{1}{2}$), shell rate h_0 of 25 Btu/(hr)(ft²)(°F) would decrease to 17.1 Btu/(hr) (ft²)(°F) (line b), tube rate h_1 of 250 would decrease to 143 (line c), pressure drop would be decreased by factor of $\frac{1}{4}$ (f_v-scale).

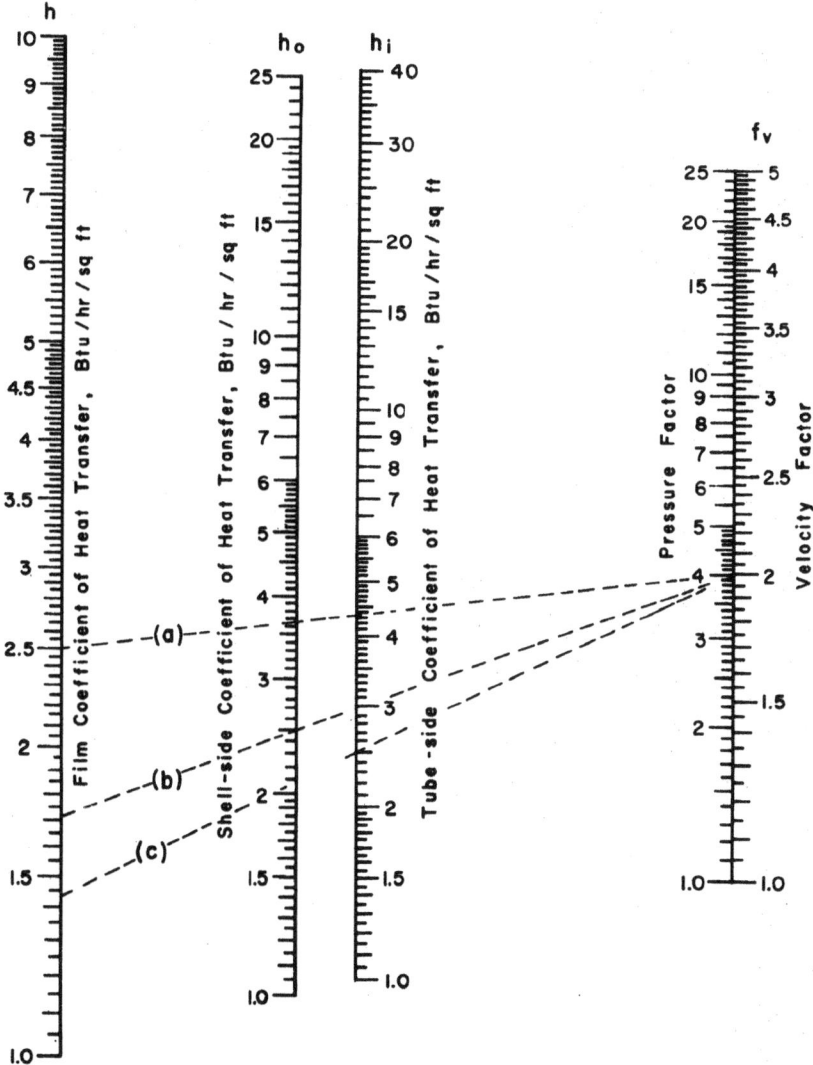

Figure 2-17

UNIT 3

Pulp and Paper Mill Calculations

Consistencies—Production—Properties

Several relatively simple nomographs deal with consistency of paper pulp; others cover rates of production of pulp mills, bleaching plants, and paper machines. General production of sheet materials, length of paper in a roll, and breaking length of paper are included.

3-1   Consistency of Paper Pulp

MALLORY D. DAVIS SR.

Pulp weight taken off a washer, the weight of the same pulp after being put through a wringer, and consistency of the pulp as it is on the washer can be correlated by the equation

$$C = KW_2/W_1$$

where   K   = wringer factor
        W_2 = wringer weight, after wringer
        W_1 = wet weight, before wringer
        C   = consistency, %

Typical Example

With a wringer factor of 30, what is the consistency of pulp if the weight of this pulp is 200 g before it is put through the wringer and 50 g after it is put through the wringer? In Figure 3-1 connect K = 30 with W_2 = 50. Connect the point of intersection on the a-scale to 200 on the W_1-scale. Intersection with the C-scale is at 7.5% consistency.

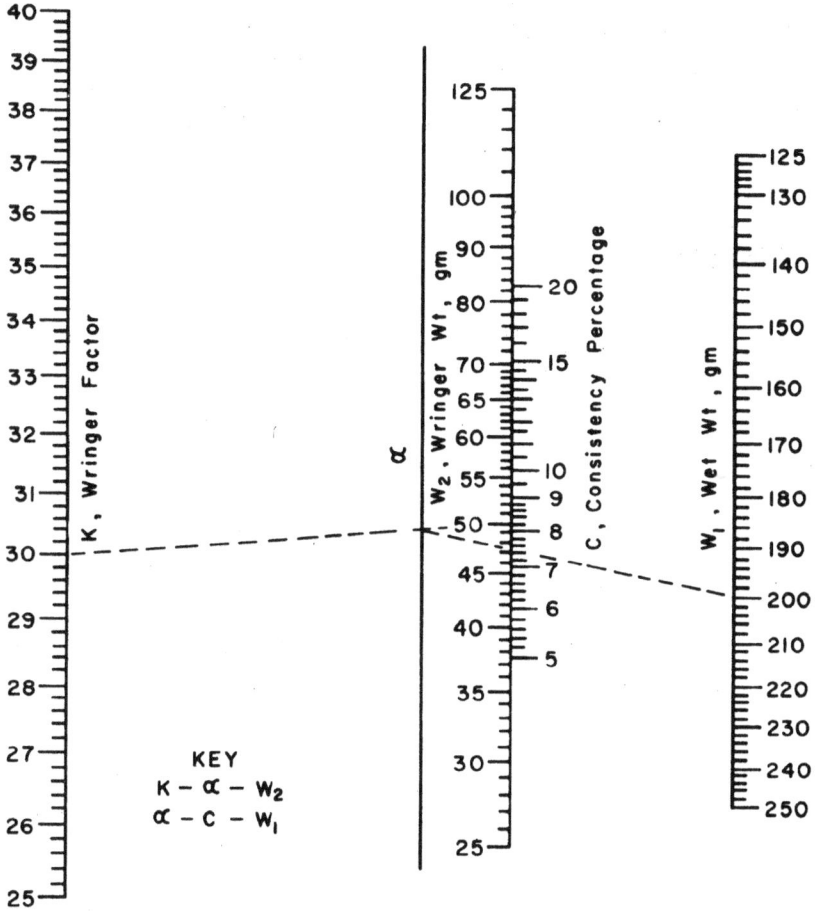

Figure 3-1

3-2   Dry Weight of Paper Pulp

S. E. HENRY

Figure 3-2 is useful in computing the dry weight of pulp contained in a given volume of pulp-water mixture at a known consistency. It may also be used for other slurries, if it may be assumed that the density of the mixture is the same as that of water.

The nomograph is based on the following equation:

$$\text{dry weight (tons)} = \text{(gallons) (percent consistency)} \ (4.17 \times 10^{-5})$$

Typical Example

Consistency   = 3.1%
Gallons   = 84,000
Tons of pulp = 10.9

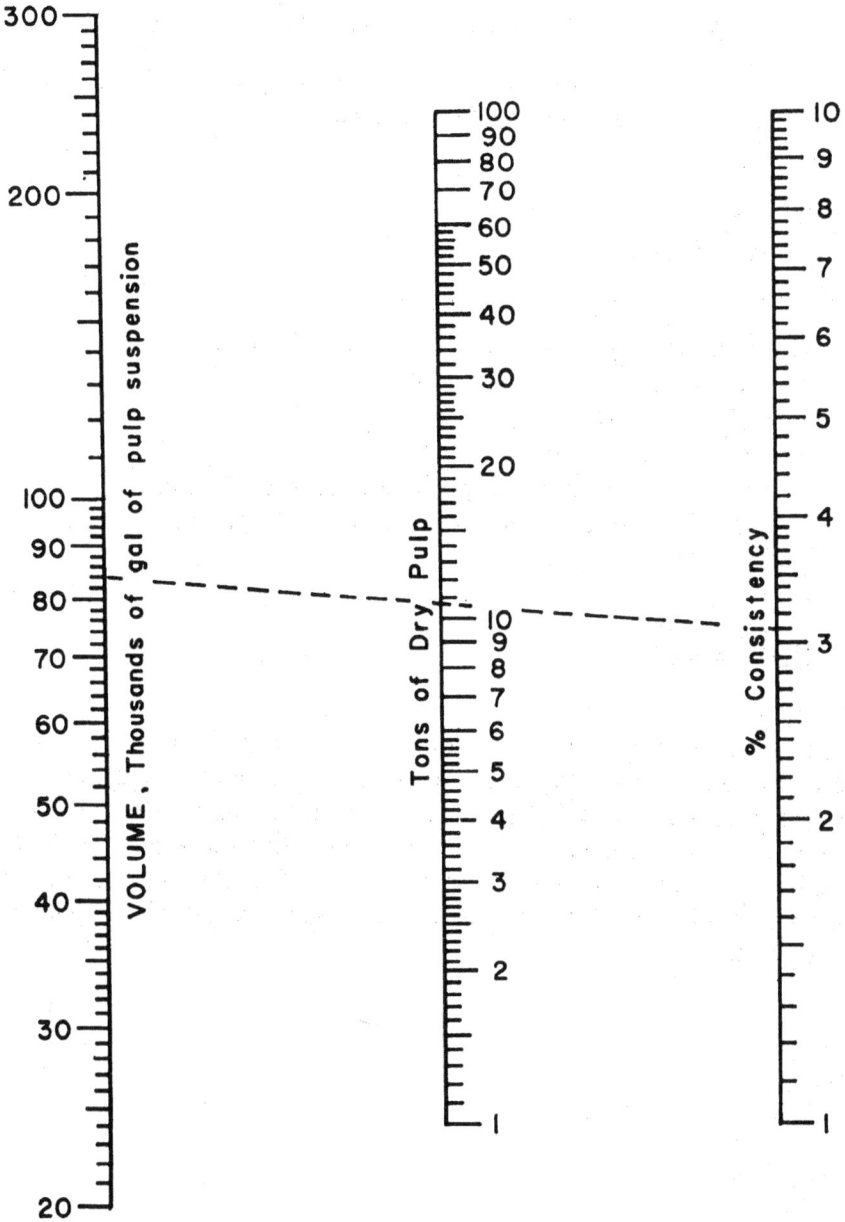

Figure 3-2

3-3   Pulp Chest Inventory

GEORGE W. LEWIS, JR.

The number of air-dry tons (10% moisture) of paper pulp in a square-type chest is given by the equation

$$T = \frac{62.5\,hWLC}{100 \times 2000 \times 0.9}$$

which simplifies to

$$T = 3.472\,hWLC \times 10^{-4}$$

where T = weight of pulp in chest, air-dry tons
        h = depth of pulp in chest, ft
       W = width of chest, ft
       L = length of chest, ft
       C = percentage oven-dry consistency of pulp
   62.5 = density of pulp of any consistency, lb/ft^3

Typical Example

For a chest width of 20 ft, chest length of 30 ft, pulp depth of 6 ft and a consistency of 3%, connect (see key) the points on the W-scale (20 ft) and the L-scale (30 ft) for the intersection on the Φ-scale. The points on the h-scale (6 ft) and the C-scale (3%) intersect on the u-scale. Connect intersection-points on the Φ-scale and the u-scale to read 3.75 tons. On the T-scale the air-dry weight of the pulp in the chest.

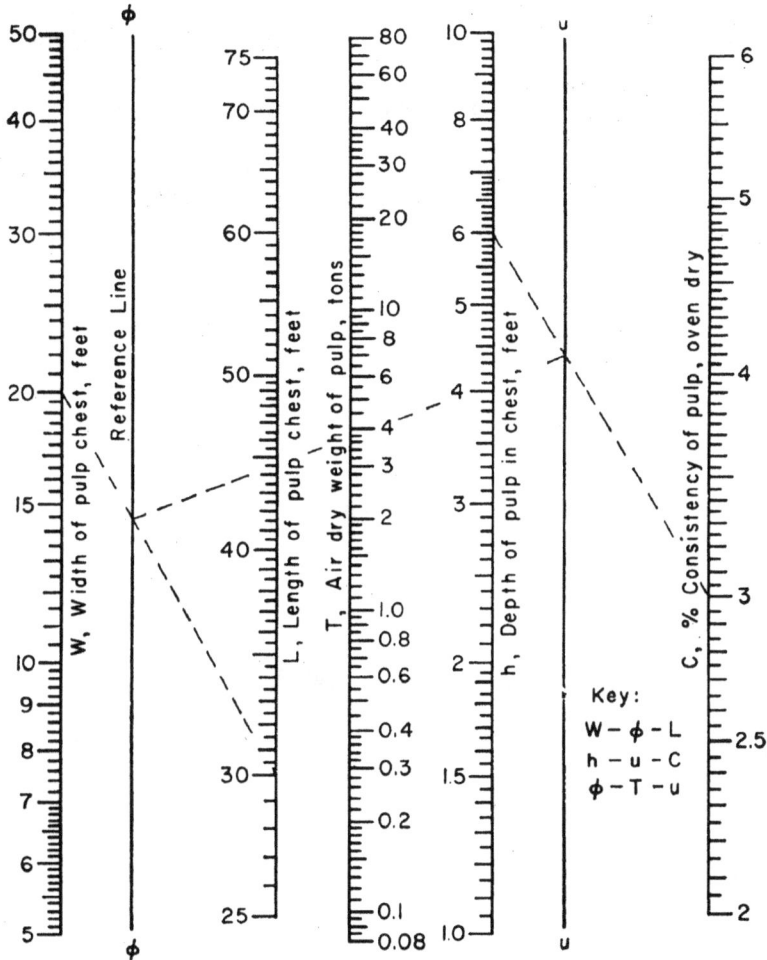

Figure 3-3

3-4   Pulp Production

S. E. HENRY

Figure 3-4 is useful in computing pulp production rate from flow rate of pulp-water mixture and the consistency. It may also be employed for slurries, if it may be assumed that the density of the mixture is the same as that of water.

The nomograph is based on the following equations:

$$\text{tons/hr} = (\text{gal/min}) \, (\% \text{ consistency}) \, (2.5 \times 10^{-3})$$
$$\text{tons/day} = (\text{gal/min}) \, (\% \text{ consistency}) \, (6 \times 10^{-2})$$

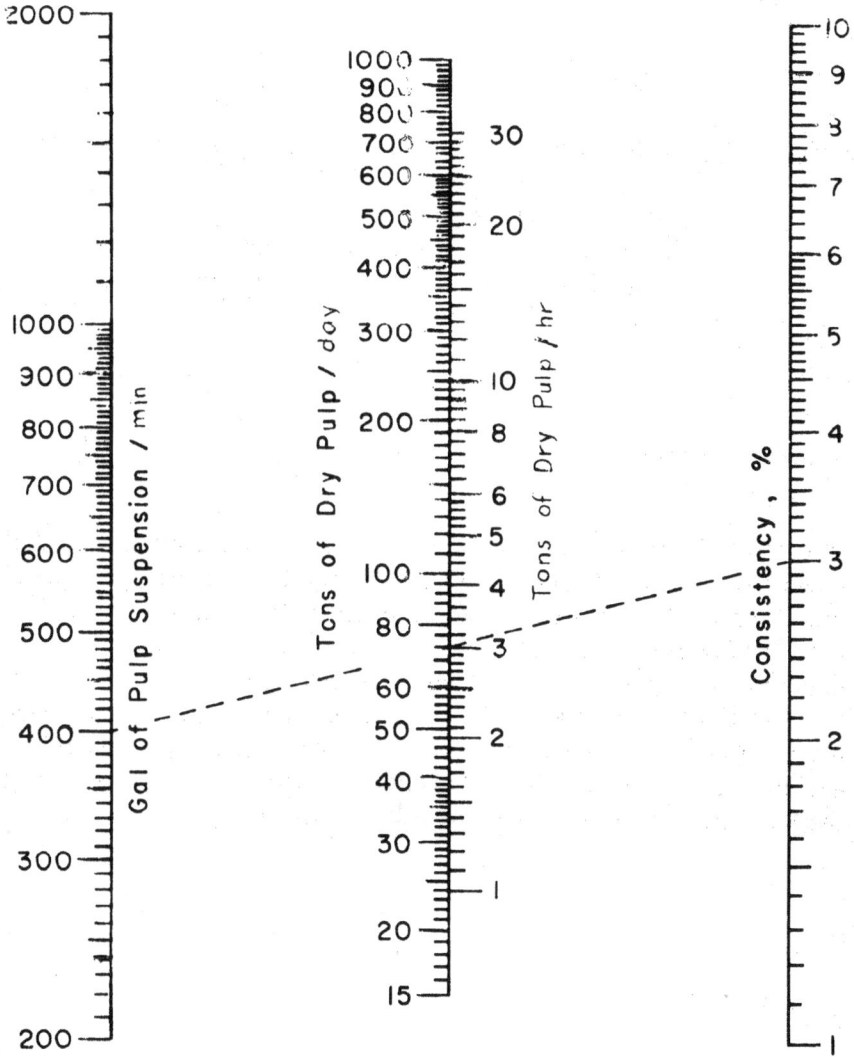

Figure 3-4

3-5   **Bleach-Plant Production**

GEORGE W. LEWIS

The tonnage passing through a paper-pulp bleaching tower is given by the equation

$$P = \frac{0.3534\ hd^2C}{R}$$

where   P = production rate, tons of oven-dry fiber per 24-hr day
      h = height of pulp in tower, ft
      d = inner diameter of tower, ft
      C = consistency of pulp on oven-dry basis, %
      R = time of retention of pulp in tower, min
      Pulp density = 62.5 lb/ft^3

Typical Example

What is the production rate of a bleaching tower 10 ft in diameter if the time required for color change to pass through the tower is 100 min, pulp consistency is 8%, and pulp level is maintained at 20 ft in tower? Connect 20 on the h-scale and 10 on the d-scale with a straight line and mark the intersection with the reference axis α_1. Connect 8 on the C-scale and 100 on the R-scale with a straight line and mark intersection with reference axis α_2. Then connect the two reference-axis intersections with straight line and read the rate of production of bleached pulp as 56.5 tons of oven-dry pulp per day.

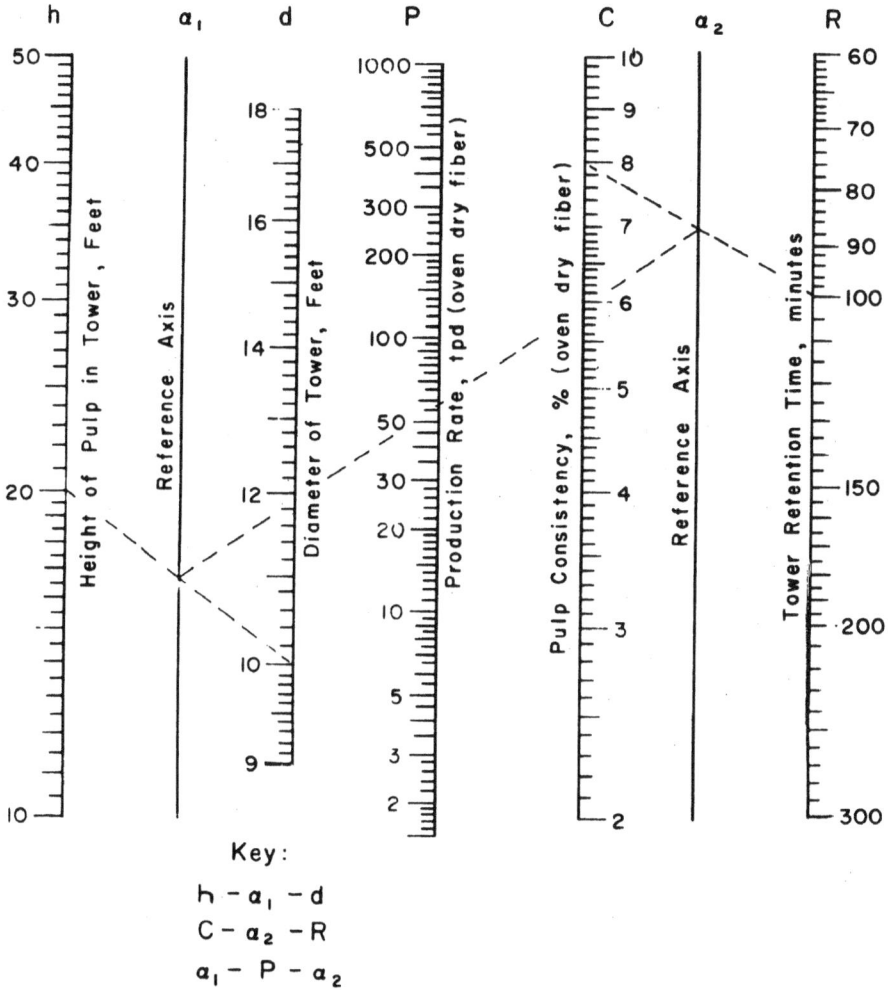

Figure 3-5

3-6   Paper-Machine Production

GEORGE W. LEWIS

Daily production of a paper machine can be determined by use of the following equation:

$$P = STW\ (2.00 \times 10^{-5})$$

where S = speed, ft/min
        T = trim, in.
        W = ream weight, 24 × 36 in., 500 sheets
        P = production, tons/24-hr day

Typical Example.

If the machine speed is 1000 ft/min with a trim of 160 in. and a ream weight of 50 lb, then the production in is 160 tons/day or 6.67 tons/hr as indicated by P_1 and P_2 respectively.

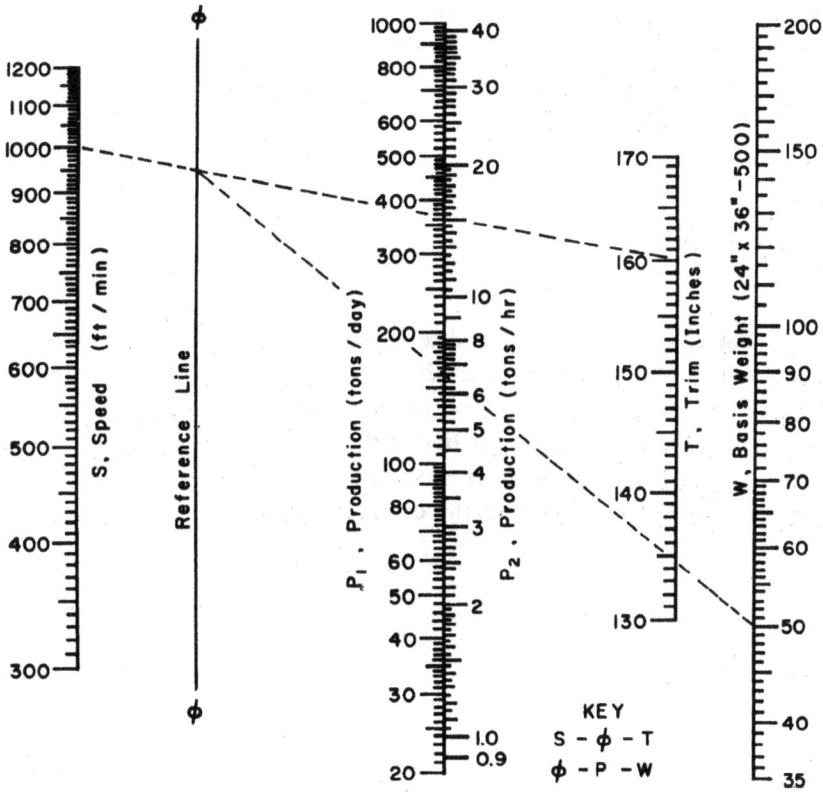

Figure 3-6

3-7   Sheet-Material Production Rate

GERALD G. WIRE

The continuous rate of production of sheet material — such as paper or paperboard — is governed by the relationship

$$P = 5\ SWw/1000$$

where P = production rate, lb/hr
       S = machine speed, ft/min
       W = sheet weight, lb/1000 ft^2
       w = sheet width in.

Typical Example

At what rate is 60 in. wide sheet material manufactured when the production machine operates at 200 ft/min and the sheet weighs 78 lb/1000 ft^2? Following the key, connect 200 on the S-scale and 78 on the W-scale with a straight line, extending it to the α-axis. Connect the reference intersection with 60 on the w-scale, and extend this line to the P-scale, where the desired value is read as 4680 lb/hr.

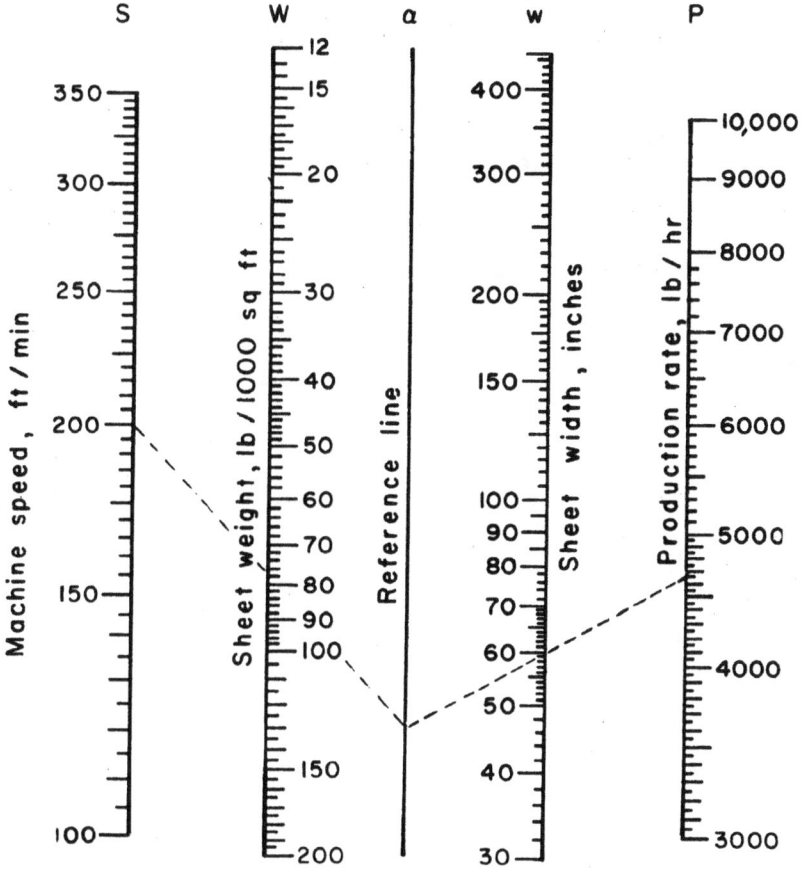

Key:
S – W – α
α – w – P

Figure 3-7

3-8   Length of Paper in a Roll

S. E. HENRY

Figure 3-8 will help to determine length of paper in a roll when caliper, roll diameter, and core diameter are known. It can prove useful when dealing with paper and paper board where differing core sizes are used. The nomograph is based on the following equation:

$$L = \frac{65.45\,(D^2 - d^2)}{C}$$

where   L   =   length of roll, ft
        C   =   caliper mils
        d   =   diameter of core, in.
        D   =   diameter of roll, in.

Typical Example

Find length of paper in roll having core diameter of 4 in. and outside diameter of 40 in., with caliper of 6 mils. Connect 40 in. on D-scale with 4 in. on d-scale. Note where this line intersects R-axis. Extend line from this point through 6 mils on C-scale until it crosses L-scale. Length of paper in this roll is 17,300 ft or 5760 yd.

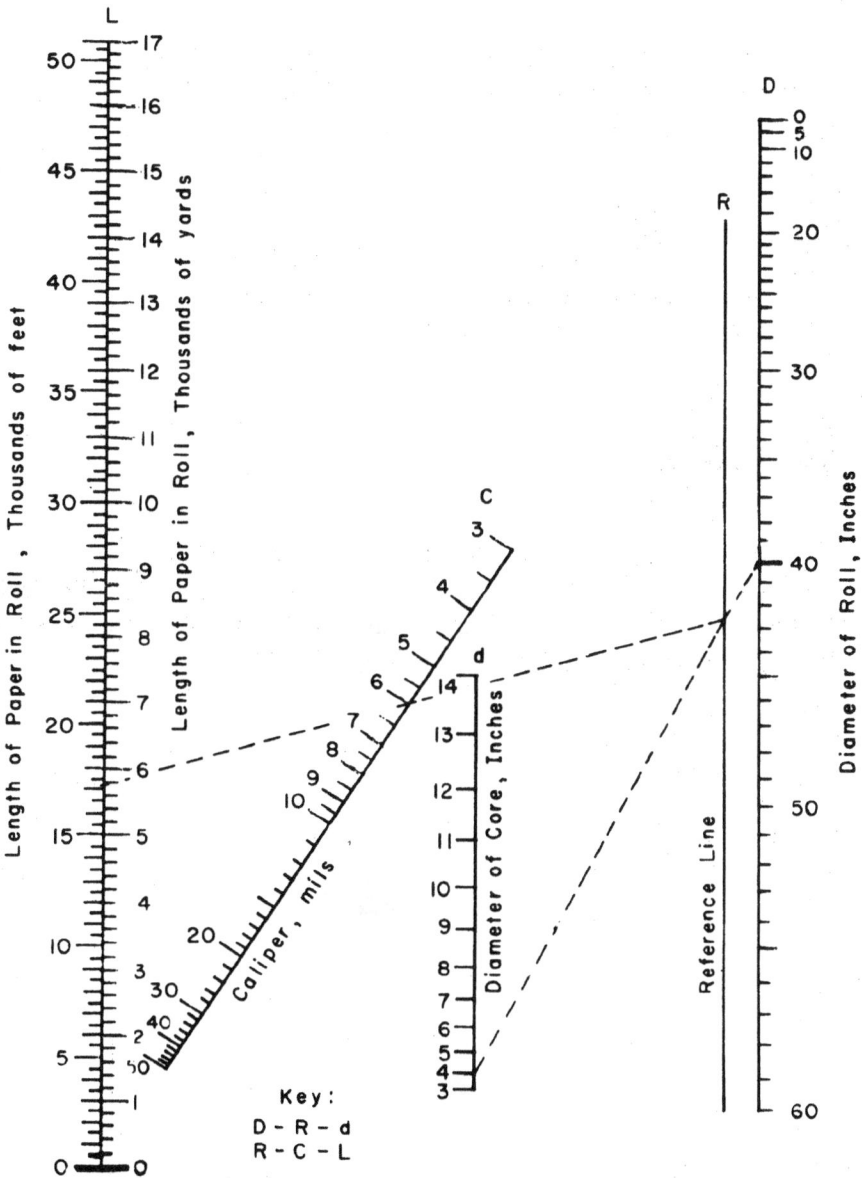

Figure 3-8

3-9   Breaking Length of Paper

GEORGE W. LEWIS

The intrinsic strength of paper as a material can be found by determining the breaking length by means of the equation

$$L = \frac{24\,tn}{W}$$

where   L   =   breaking length, yd
        t   =   tensile break load, lb (for a 1-in. strip)
        n   =   number of 24 × 36-in. sheets/ream
       W   =   weight of ream, lb

Typical Example

Paper with ream weight of 50 lb (24 × 36 in., 500 sheets) exhibits a tensile break load of 34 lb. What is the breaking length of the paper? On Figure 3-9 dashed index lines, drawn in accordance with the key, show the breaking length to be roughly 8200 yd.

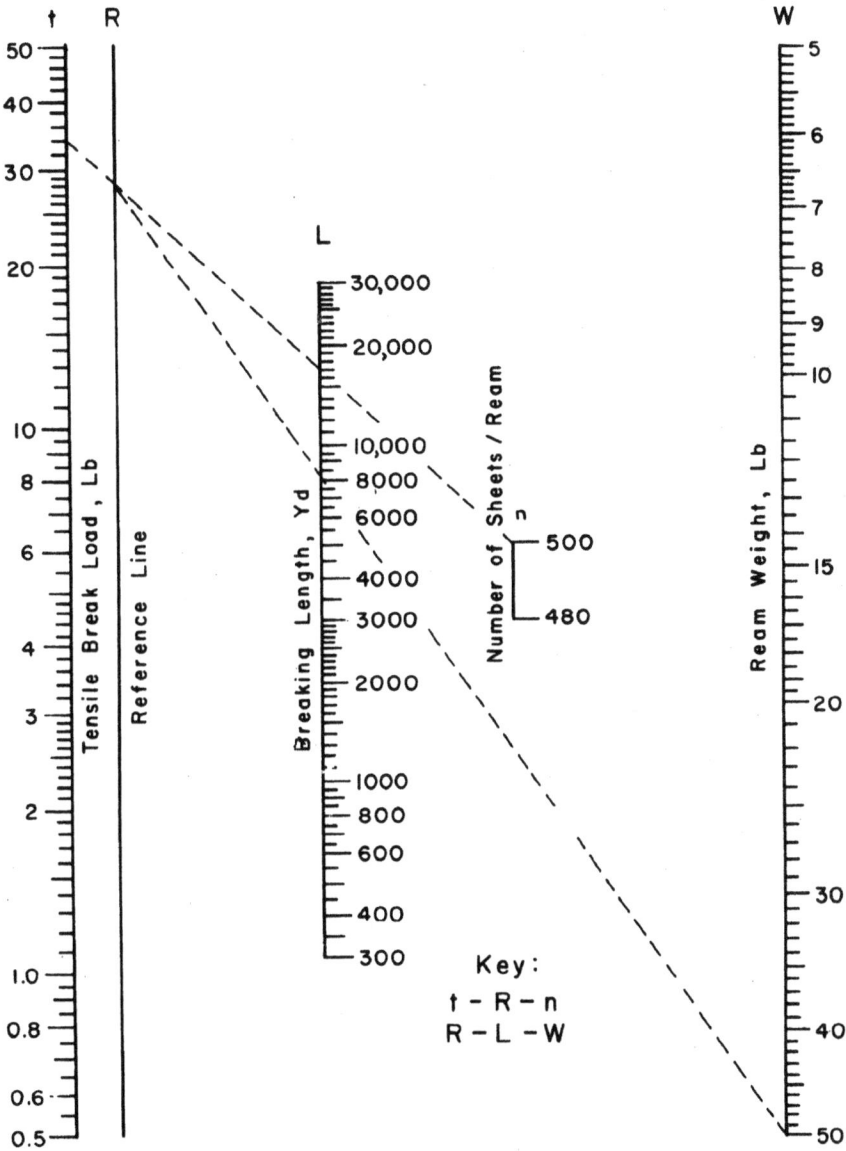

Figure 3-9

UNIT 4

Vapor Pressures

Hydrocarbons—Liquefied Petroleum Gases—Other Organic Compounds

Vapor pressures of a wide variety of organic compounds are presented in ten convenient charts. Many of them consist of temperature and pressure scales with gage points specific to the compounds and ranges of temperature. Each chart extends considerably the utility of the original data.

4-1   Vapor Pressures of Alkene Hydrocarbons

BASIL C. DOUMAS

Figure 4-1 presents a correlation between temperature and vapor pressure for 16 alkene hydrocarbon compounds in the C_7 and C_8 range. Key letters used in the chart refer to specific compounds listed in the accompanying table.[1] The nomograph is based on the Antoine equation:

$$\log_{10} P = A - \frac{B}{C+t}$$

where   P   = vapor pressure, mm Hg
     t   = temperature, °C
  A, B, C   = constants, depending on the specific compound

Typical Example

What is the vapor pressure of 4,4-dimethyl-cis-2-pentene at 80°C? Connect 80 on the F temperature scale with the point labeled by code letter F. Extend this line until it intersects the P-scale, and read the desired value, 750 mm Hg.

Key to Alkene Hydrocarbons on Nomograph

Code	Compound
A	3-Methyl-cis-3-hexene
B	3-Methyl-trans-3-hexene
C	2,4-Dimethyl-l-pentene
D	4,4-Dimethyl-1-pentene
E	2,4-Dimethyl-2-pentene
F	4,4-Dimethyl-cis-2-pentene
G	4,4-Dimethyl-trans-2-pentene
H	3-Methyl-2-ethyl-1-butene
J	2,3,3-Trimethyl-1-butene
K	2,2-Dimethyl-cis-3-hexene
L	2,2-Dimethyl-trans-3-hexene
M	2-Methyl-3-ethyl-1-pentene
N	2,4,4-Trimethyl-1-pentene
O	2,4,4-Trimethyl-2-pentene
P	1-Methylcyclohexene
R	1-Ethylcyclohexene

[1]Camin, D. L. and F. D. Rossini, *J. Chem. Eng. Data,* **5**, 368, 1960.

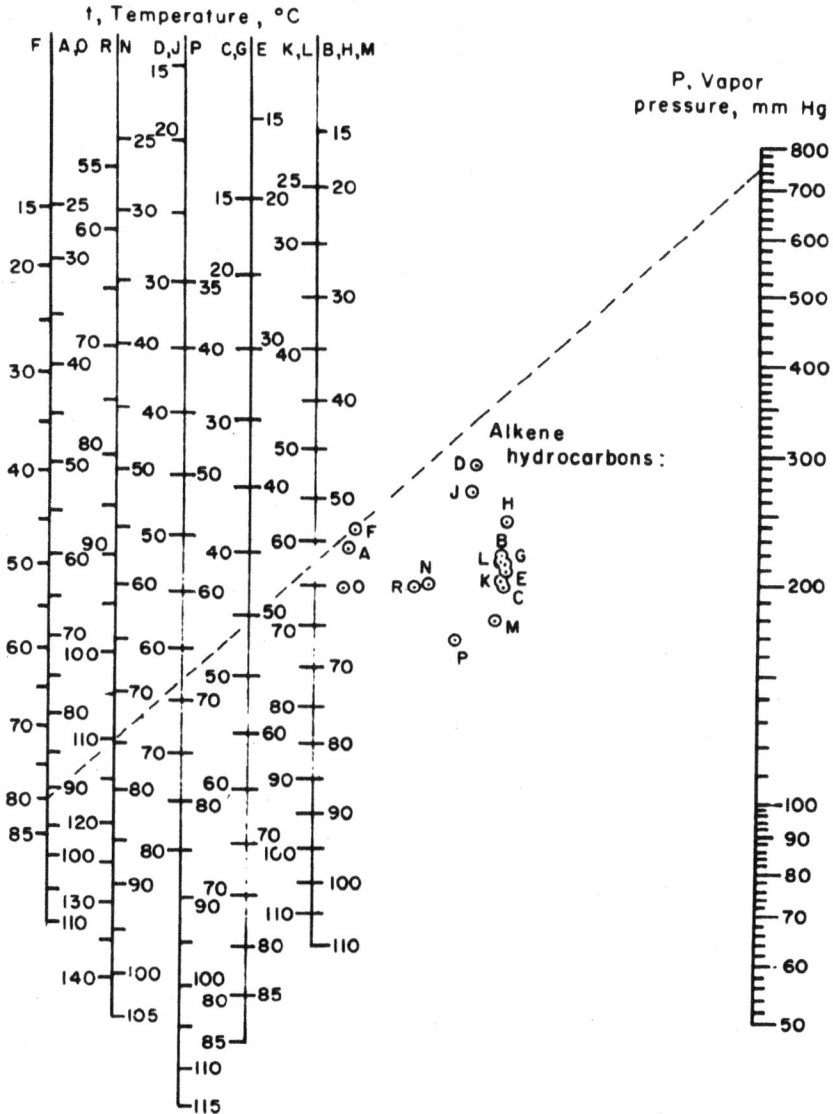

t, Temperature , °C

P, Vapor
pressure, mm Hg

Alkene
hydrocarbons :

Figure 4-1

4-2   Vapor Pressures of Normal Alkanes

JOHN B. GARDNER

The broken index line on the accompanying nomograph, Figure 4-2, shows that the vapor pressure of N-heptane is 40 mm Hg at 22°C. The points refer to the number of carbon atoms for normal alkane series.

The chart agrees well with tables on pages 153-165 of the third edition of Perry's *Chemical Engineers' Handbook*.

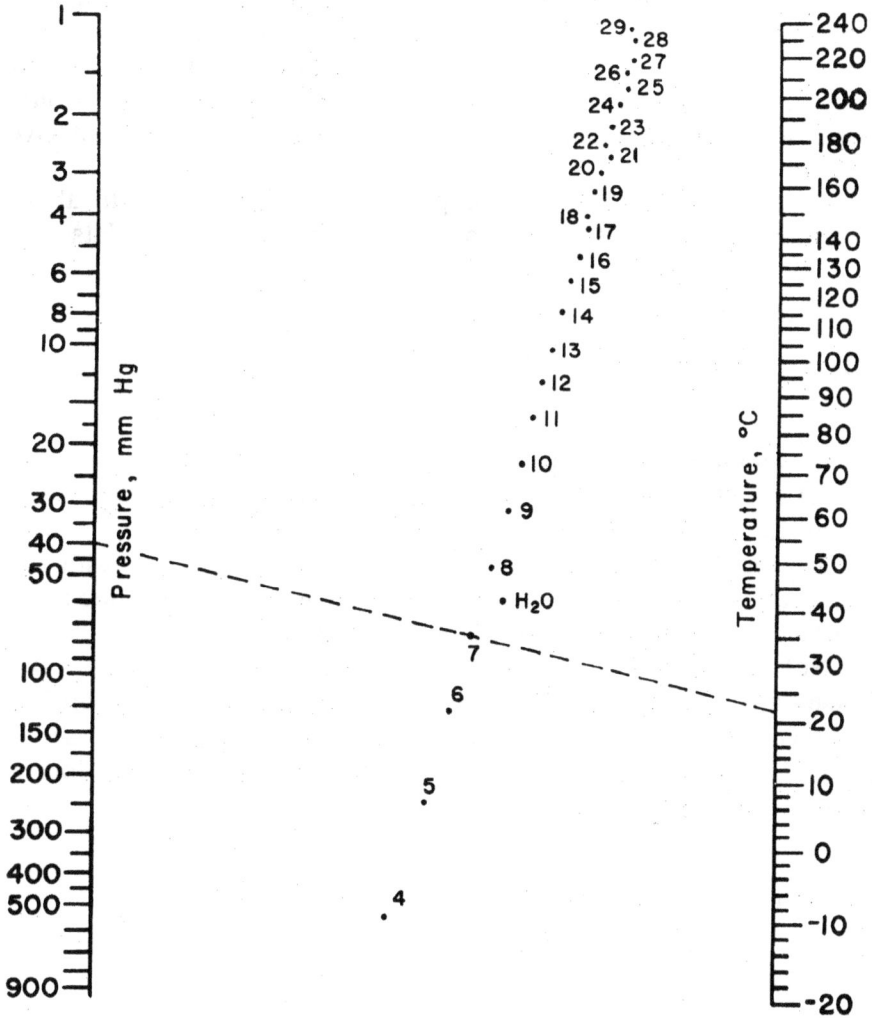

Figure 4-2

4-3 Vapor Pressure of Natural Gasoline and Liquefied Petroleum Gases

GEORGE E. MAPSTONE

Information on pressure changes with temperature is needed in planning storage for high vapor-pressure products such as natural gasoline or liquefied petroleum gases. As effective molal latent heats of vaporization are, for these mixtures, much lower than that of pure hydrocarbons, interpolation of pure hydrocarbon data will not give accurate results.

Data[1] are available for vapor pressures of gasolines up to about 45 lb/in.² Figure 4-3 is based on an extrapolation of these data.

Typical Example

A certain mixture of petroleum products shows a vapor pressure of 24 lb/in.² abs. at 68°F. What vapor pressure will be reached if the temperature rises to 89°F?

Connect 24 lb/in.² abs. and 68°F. Using the intersection with the tie line as a pivot, swing the straight-edge down the temperature scale to 89°F. The vapor pressure is 32 lb/in.² abs.

Note that points are given for pure hydrocarbons. Where the material in question is largely one of these, with some impurities, neither tie line nor point will give accurate results. In such cases it is best to measure the vapor pressure at two temperatures and locate a point appropriate for the product.

[1]Mapstone, G. E., *Pet. Ref.*, **30**, (10), pp. 157-9, 1951.

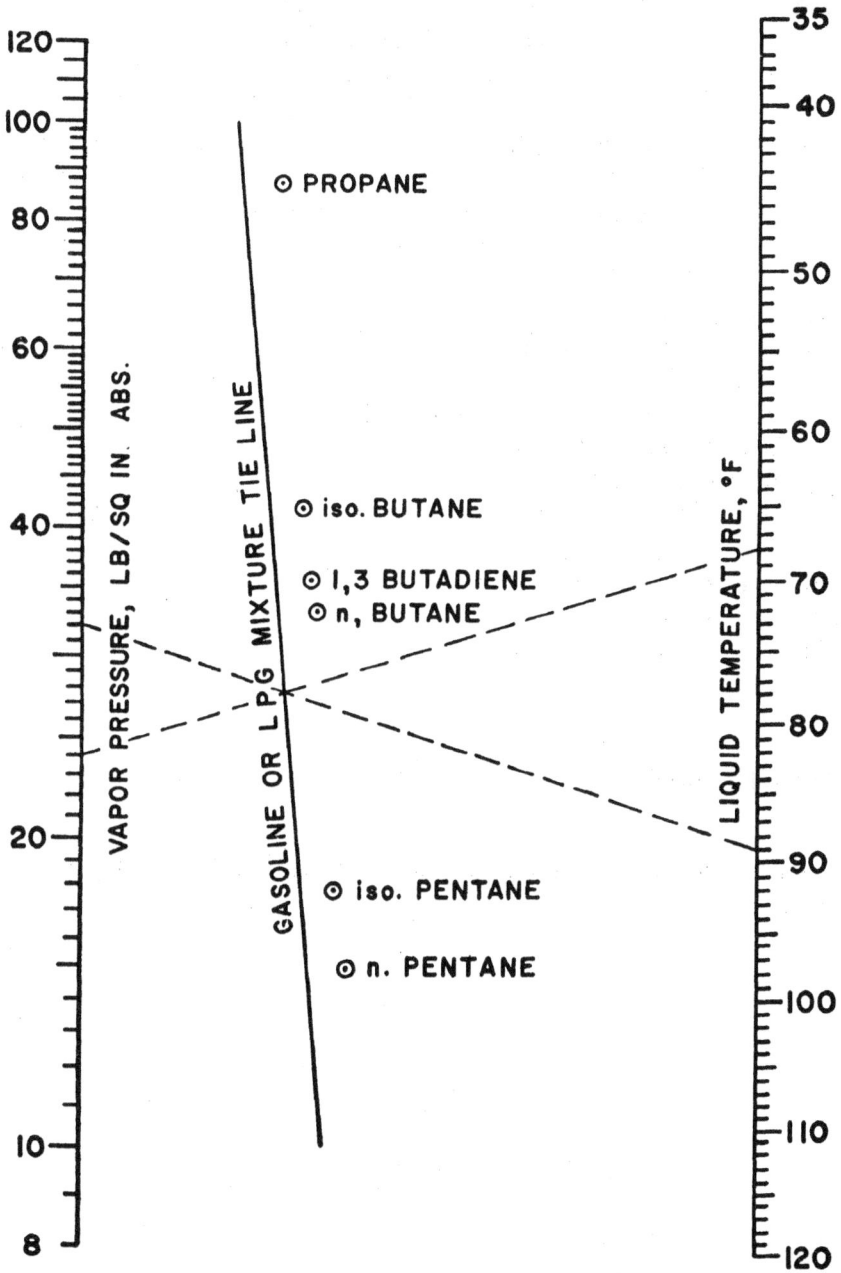

Figure 4-3

4-4   Vapor Pressures of Chloromethanes and Chloroethanes

JOHN B. GARDNER

The vapor pressures of the listed compounds can be found by drawing a straight line from the desired temperature, through the number of the compound in question, and to the pressure line in Figure 4-4.

1) Hexachloroethane, C_2Cl_6
2) Pentachloroethane, C_2HCl_5
3) 1,1,2,2 Tetrachloroethane, $C_2H_2Cl_4$
4) 1,1,1,2 Tetrachloroethane, $C_2H_2Cl_4$
5) 1,1,2 Trichloroethane, $C_2H_2Cl_3$
6) 1,2 Dichloroethane, $C_2H_4Cl_2$
7) Tetrachloromethane, CCl_4
8) 1,1,1 Trichloroethane, $C_2H_3Cl_3$
9) Trichloromethane, $CHCl_3$
10) 1,1 Dichloroethane, $C_2H_4Cl_2$
11) Dichloromethane, CH_2Cl_2
12) Chloroethane, C_2H_5Cl

Typical Example

The vapor pressure of carbon tetrachloride at 23°C is 100 mm Hg.

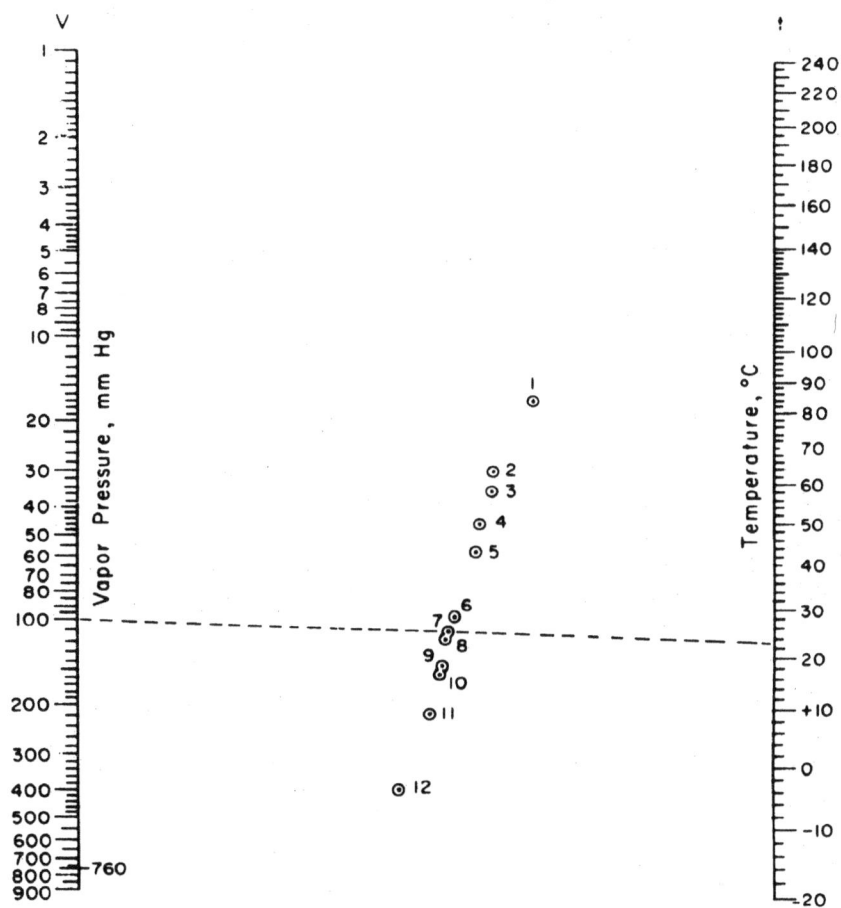

Figure 4-4

4-5   **Vapor Pressures of Chlorinated
Ethylenes**

JOHN B. GARDNER

To use Figure 4-5 for chlorinated ethylenes, draw straight lines
through numbered gage points to find vapor pressures at any temperature. Also, draw straight lines through numbered gage points to
find boiling points at any pressure.

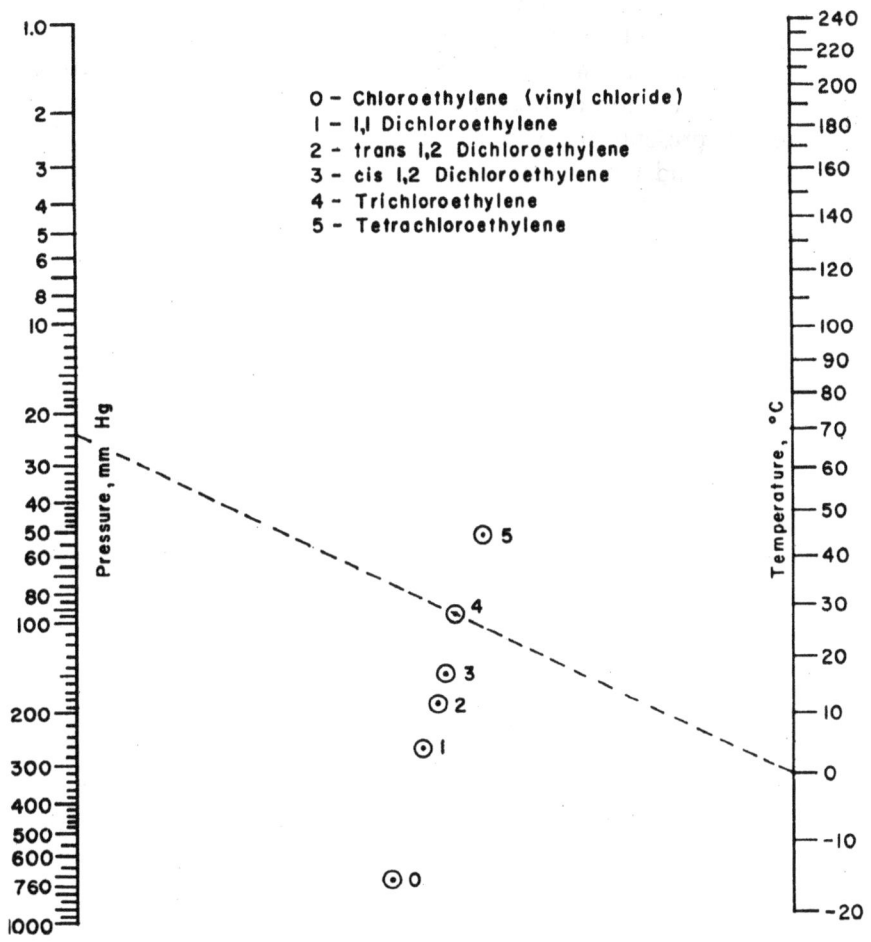

Figure 4-5

4-6   Vapor Pressures of Wet Diethylene Glycols

GEORGE E. MAPSTONE

Diethylene glycol is frequently used to dehumidify gases because of its low vapor pressure and high water-holding capacity.

To design a plant for this purpose, it is necessary to know the vapor pressure of the wet glycol both at the absorber temperature and at the temperature of the regenerator still.

Figure 4-6 presents the vapor-pressure data as a function of the water content and temperature of the glycol.

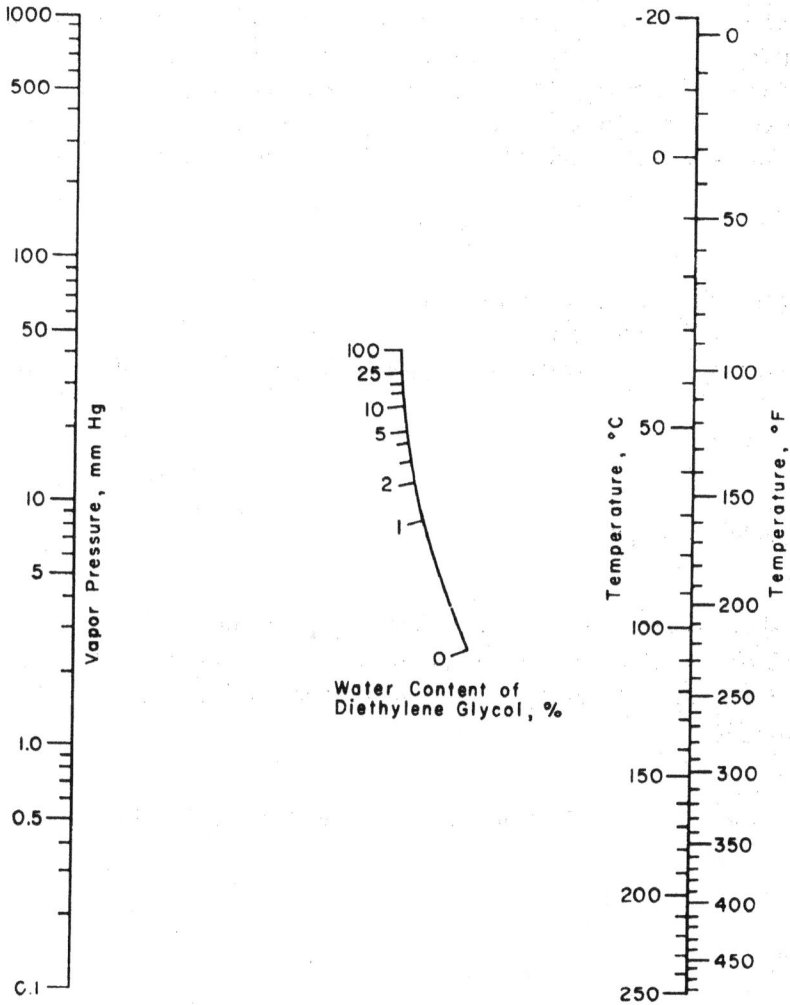

Water Content of
Diethylene Glycol, %

Figure 4-6

4-7 Vapor Pressures of Methyl Esters of Common Saturated Normal Fatty Acids

BASIL C. DOUMAS

Figure 4-7 presents a correlation between temperature and vapor pressure for seven methyl esters of the normal saturated fatty acids having 6, 8, 10, 12, 14, 16 and 18 carbon atoms, in the range between 20 and 200 mm Hg absolute pressure. Key letters in the chart refer to specific compounds listed in the accompanying table. The nomograph is based on reliable data[1] and the Antoine equation:

$$\log_{10} P = A - \frac{B}{C+T}$$

where P = vapor pressure, mm Hg
       T = temperature, °C
   A, B, C = constants, values depending on the specific compound

Typical Example

What is the vapor pressure of methyl stearate at 222°C? Connect 222 on the G-scale with the point labeled by code letter G. Extend this line until it intersects the P-scale, and read the desired value, 19 mm Hg.

[1]Rose, A. and W. R. Supina, *J. Chem. Eng. Data*, **6**, 173, 1961.

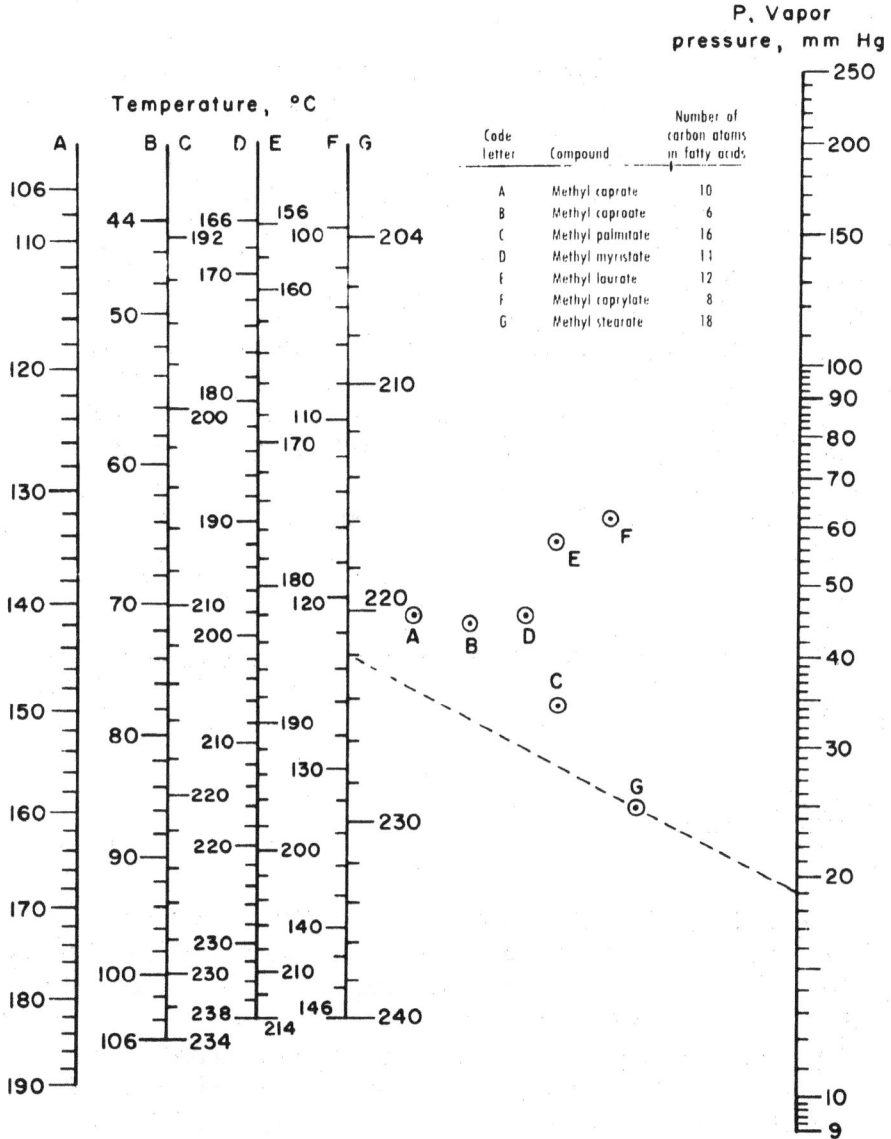

Figure 4-7

4-8   Vapor Pressure of Methyl Esters

D. S. DAVIS

Scott, MacMillan, and Melvin[1] determined the vapor pressures of methyl esters of some fatty acids and correlated the pressure-temperature data by means of the Antoine equation:

$$\log p = A - \frac{B}{t+C}$$

where   P   = pressure, mm of mercury
         t   = temperature, °C
A, B, and C depend on the compound (the value of C varying from 113 to 226).

As the original data are given only in tabular form, there is evident need for a convenient graphical presentation, so that vapor pressures can be estimated at any desired temperature in the ranges of the investigation.   Figure 4-8 is based on the equation

$$\log p = A - \frac{B}{t+160}$$

where a value of C = 160, for all the compounds, replaces individual values.

Typical Examples

Use of the chart is illustrated as follows: What is the vapor pressure of methyl laurate at 100°C? A straight line drawn from 100 on the t-scale to point 3, for methyl laurate, intersects the p-scale at the desired value, 1.65 mm of mercury.

At what temperature is the vapor pressure of methyl caprate 4.0 mm of mercury? A straight line drawn from 4.0 on the p-scale to point 2, for methyl caprate, intersects the t-scale at 89°C, the desired value.

[1]Scott, T. A., Jr. Macmillan, D., and Melvin, E. H., *Ind. Eng. Chem.*, **44**, p. 172, 1952.

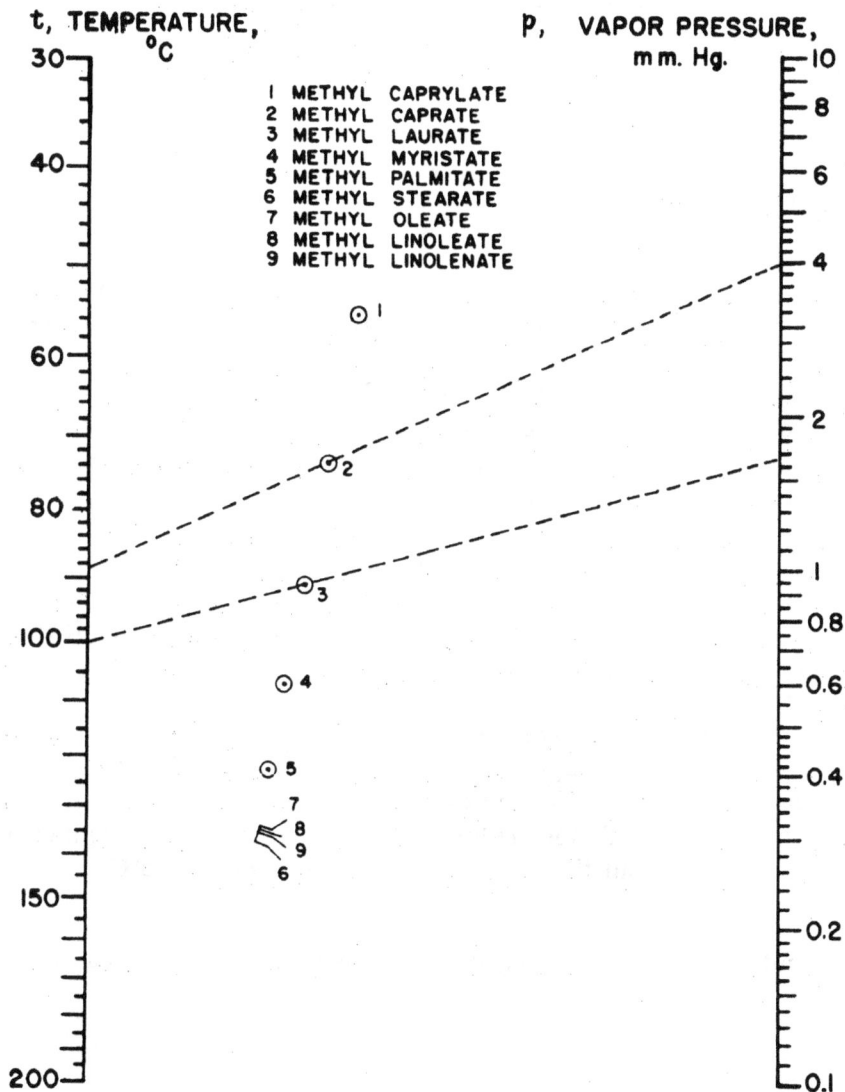

t, TEMPERATURE, °C

P, VAPOR PRESSURE, mm. Hg.

1 METHYL CAPRYLATE
2 METHYL CAPRATE
3 METHYL LAURATE
4 METHYL MYRISTATE
5 METHYL PALMITATE
6 METHYL STEARATE
7 METHYL OLEATE
8 METHYL LINOLEATE
9 METHYL LINOLENATE

Figure 4-8

4-9   Vapor Pressure Data

BRUCE FADER and FRANK McELROY

The vapor pressures of various compounds[1] can be found by applying the listed coordinates of Table 4-1 to Figure 4-9. Note that in many cases, the coordinates themselves are a function of temperature.

Typical Examples

A certain compound is listed with coordinates $X = 11$, $Y = 61$. Its vapor pressure at a given temperature is found by extending a straight line drawn from this temperature on the right scale, through the coordinates. In this case, a 45°C temperature gives a vapor pressure of about 1520 mm Hg. At what temperature will the compound boil at standard pressure? Connect 760 mm with the coordinates and extend the line. Read 20°C.

The nomograph can also be used to approximate the vapor pressure of paraffin mixtures. For any mixture of paraffins where the vapor pressure is known for some temperature, connect the known temperature and the vapor pressure and note the point on the tie-line. Pivot about this point to find the vapor pressure of the mixture at other temperatures.

A mixture of light paraffins is known to boil at 20°C. At what temperature will the vapor pressure reach two atmospheres? Connect 20°C and 760 mm as the known boiling condition. Locate the intersection with the tie-line. Pivoting, move the straight-edge to 1520 mm. Read the temperature (approximate) as 45°C.

---

[1]Jordan, T. E., *Vapor Pressure of Organic Compounds*, Interscience Publishers, New York, 1954.

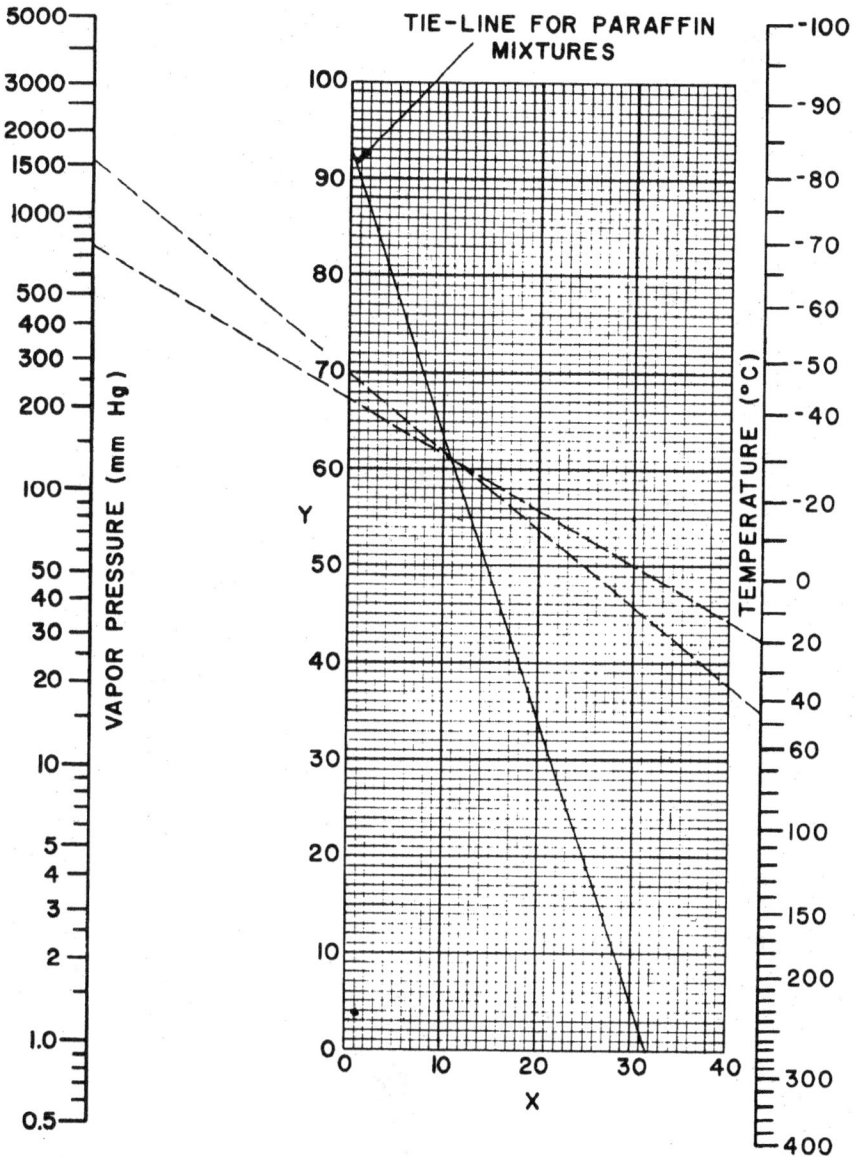

Figure 4-9

Table 4-1
Temperature Ranges and Coordinates for Use with Figure 4-9

Compound	Temperature Range, °C	Coordinates	
		X	Y
Acetone	−70 to 125	18.0	50.0
Acetylene	— —	12.3	88.8
Amyl acetate	0 to 155	23.5	34.0
Anisole	5 to 156	23.0	33.5
Benzene	below 25	19.4	46.3
	above 25	18.0	46.8
1,2-Butadiene	— —	14.1	59.2
1,3-Butadiene	below −80	15.1	65.7
	−80 to −35	14.0	64.4
	above −35	12.6	64.7
Butane	below −68	14.8	64.3
	above −68	13.4	63.7
Butanol	0 to 185	24.0	36.5
Butyl acetate	0 to 135	22.5	37.5
Butylbenzene	below 60	24.7	30.4
	above 60	23.3	30.5
1-Butyne	— —	15.0	61.0
Carbon disulfide	−82 to 155	16.0	53.5
Carbon tetrachloride	0 to 150	19.0	46.5
Chloroform	−60 to 61	19.0	49.5
	61 to 130	17.5	50.0
Cumene	below 40	23.8	34.8
	40 to 90	22.2	35.7
	above 90	21.4	35.4
Cyclobutane	— —	14.4	60.6
Cyclohexene	below −20	21.2	47.0
	−20 to 15	20.5	46.5
	above 15	18.6	46.5
Cyclopropane	below −77	14.0	72.9
	above −77	11.9	71.9
p-Cymene	0 to 177	23.5	31.5
Decane	below 55	24.1	31.6
	above 55	23.2	31.8
Diacetone alcohol	22 to 178	25.0	30.5
Diethanolamine	107 to 271	29.0	17.5
Diethylene glycol	80 to 255	28.5	19.5
1,1-Dimethylcyclohexane	below 60	21.4	39.9
	above 60	20.0	40.5
2,2-Dimethylhexane	below 50	20.3	42.0
	above 50	19.6	42.1
3,3-Dimethylhexane	below 52	21.6	40.9
	above 52	19.9	41.3
Dipropylene glycol	74 to 240	27.5	22.0
Ethane	— —	5.4	89.5
Ethanol	−38 to 135	22.0	43.0
Ether	−85 to 102	16.5	54.5
Ethyl acetate	−43 to 145	19.5	46.0
Ethyl benzene	below 75	22.2	37.0
	above 75	21.1	37.3
Ethylcyclopentane	below 45	20.4	42.5
	above 45	19.2	42.7
Ethylene glycol	45 to 205	27.0	25.5
3-Ethyltoluene	below 45	24.1	33.7
	45 to 95	23.3	33.6
	above 95	21.9	34.0
Formaldehyde	−100 to −13	67.5	13.5
Glycerol	125 to 300	29.5	15.0

Table 4-1 (*Cont.*)

Compound	Temderature Range, °C	Coordinates X	Y
Heptane	below 38	21.3	43.0
	above 38	19.5	43.4
1,5-Hexadiene-3-yne	— —	18.2	46.0
Hexane	below 15	18.6	48.6
	above 15	17.8	48.8
1-Hexene	below —30	18.8	49.5
	above —30	18.1	48.8
Isobutanol	—10 to 165	22.5	38.5
Isopropanol	—25 to 140	23.0	42.0
Mesityl oxide	0 to 137	22.5	37.0
Methanol	—44 to 123	21.0	46.5
Methyl acetate	—57 to 125	19.0	49.5
Methyl amyl ketone	10 to 160	25.0	32.0
3-Methyl-1-butene	— —	15.1	58.9
Methyl chloride	—70 to 33	10.5	69.5
Methylcyclohexane	below 40	19.9	43.1
	above 40	19.6	43.2
Methylene chloride	—70 to 41	17.5	53.0
2-Methylhexane	below 35	20.2	44.3
	above 35	18.7	44.8
Methyl ethyl ketone	—48 to 90	19.0	46.0
2-Methyl-2-heptane	below 68	22.1	38.8
	aboe 68	21.3	39.0
Nitromethane	—28 to 110	21.0	41.5
1-Nitropropane	0 to 143	22.5	37.0
Octacosane	— —	32.6	4.1
Octadecane	below 240	28.0	15.3
	above 240	27.7	15.8
Octane	— —	22.0	38.3
Pentane	below —10	16.6	55.2
	above —10	15.5	55.4
1-Pentene	below —17	16.0	56.3
	above —17	15.3	56.5
Perchloroethylene	—20 to 130	21.0	39.5
Propadiene	below —73	12.0	73.0
	above —73	11.0	72.3
Propane	below —80	11.0	75.3
	above —80	9.4	74.5
Propanol	—15 to 158	24.0	39.5
Propyl acetate	—35 to 165	21.5	41.5
Propylene	— —	8.3	75.7
Propylene dichloride	0 to 105	19.0	43.5
1,2-Propylene glycol	50 to 195	25.0	28.0
Styrene	— —	21.2	36.1
Tetradecane	below 165	26.9	21.6
	above 165	25.9	22.0
Toluene	below 55	21.8	40.9
	above 55	20.0	41.7
Trichloroethylene	—20 to 100	19.5	45.0
1,2,3-Trimethylbenzene	below 60	24.2	31.3
	above 60	23.2	31.5
1,2,4-Trimethylbenzene	below 50	24.7	32.6
	50 to 100	23.0	32.4
	above 100	22.8	32.6
p-Xylene	below 80	22.7	36.5
	above 80	21.2	37.1

4-10   Vapor Pressures of Mixtures of Propylene Oxide and Propoxylated Glycerine

C. A. PLANK and C. W. YOST

Commercial preparation of polypropylene glycol compounds involves reaction of propylene oxide with glycerine and higher weight triols. Analysis of the reaction mixture is time consuming and cumbersome.

Data from our laboratories have shown that the vapor pressures of mixtures of propylene oxide and propoxylated glycerine are essentially independent of molecular weights (266-3000) and catalyst concentrations. Analysis is therefore possible by temperature/pressure relationships.

Typical Example

A mixture of PO-triol has been added to a batch reactor. The temperature is 240°F and the pressure is 60 lb/in.² abs. What is the composition of the liquid mixture? In Figure 4-10 the dashed line connecting the temperature and pressure shows the mixture to be 20% propylene oxide.

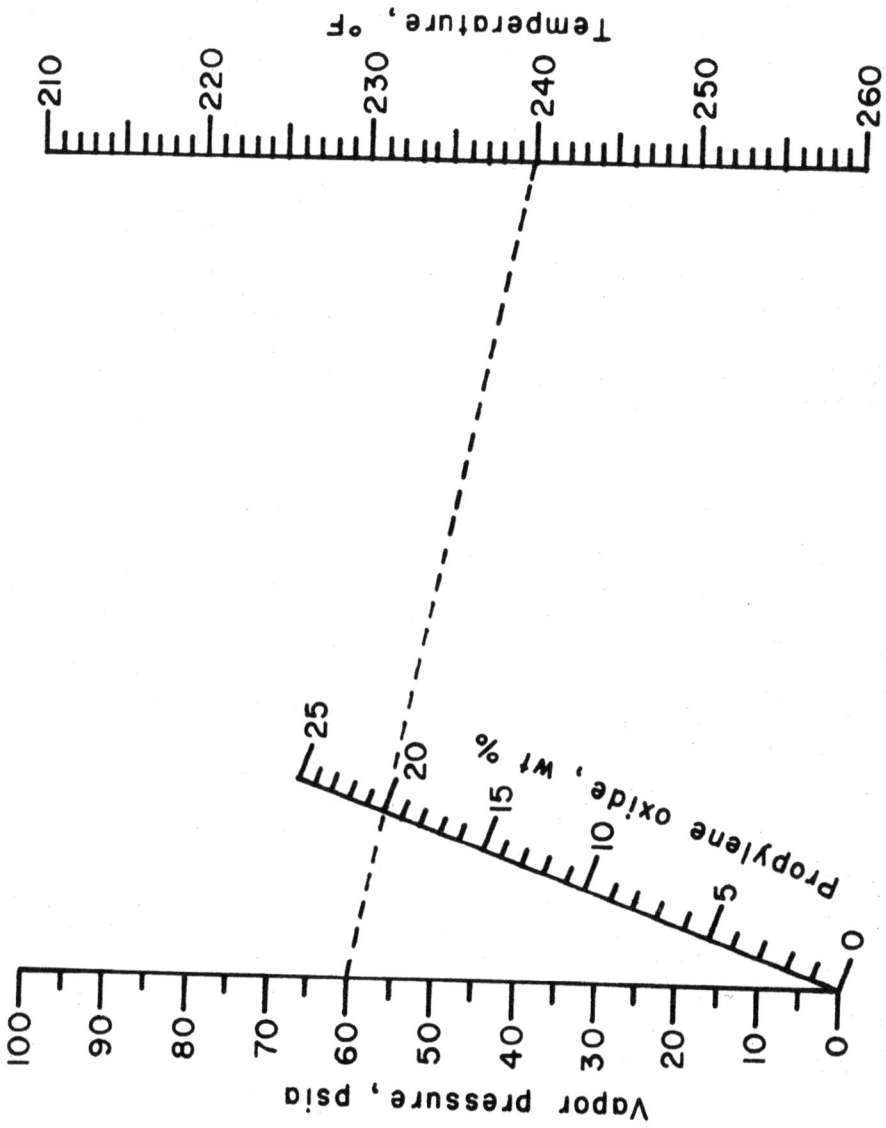

Figure 4-10

UNIT 5

Densities and Specific Gravities

Fatty Acids—Ketones—Alcohols—Mineral Acids—Asphaltic Materials

Densities or specific gravities of fatty acids, commercial formalin solutions, ketones, aqueous acetone, aqueous alcohols, two mineral acids, and heavy hydrocarbon fractions over wide ranges of temperature are given in a series of carefully designed nomographs. The charts are based on reliable data from a number of sources.

5-1 Densities of Molten Fatty Acids and Lower Oleic Alkyl Esters

GEORGE E. MAPSTONE

Published data[1,2] on the densities of the fatty acids above their melting points and for the lower oleic alkyl esters are presented in convenient nomographic form in Figure 5-1. Melting points of the acids are also indicated on the temperature scale for reference and for convenience in insuring that the acid is liquid at the temperature required.

Typical Example

The dotted line shows that lauric acid has a density of 0.869 g/cm³ at 50°C.

[1]Markley, K. S., *Fatty Acids*, Interscience Publishers, New York, 1947.
[2]Keffler, L., and McLean, J. H., *J. Soc. Chem. Ind.*, **54**, 178-185T, 1935.

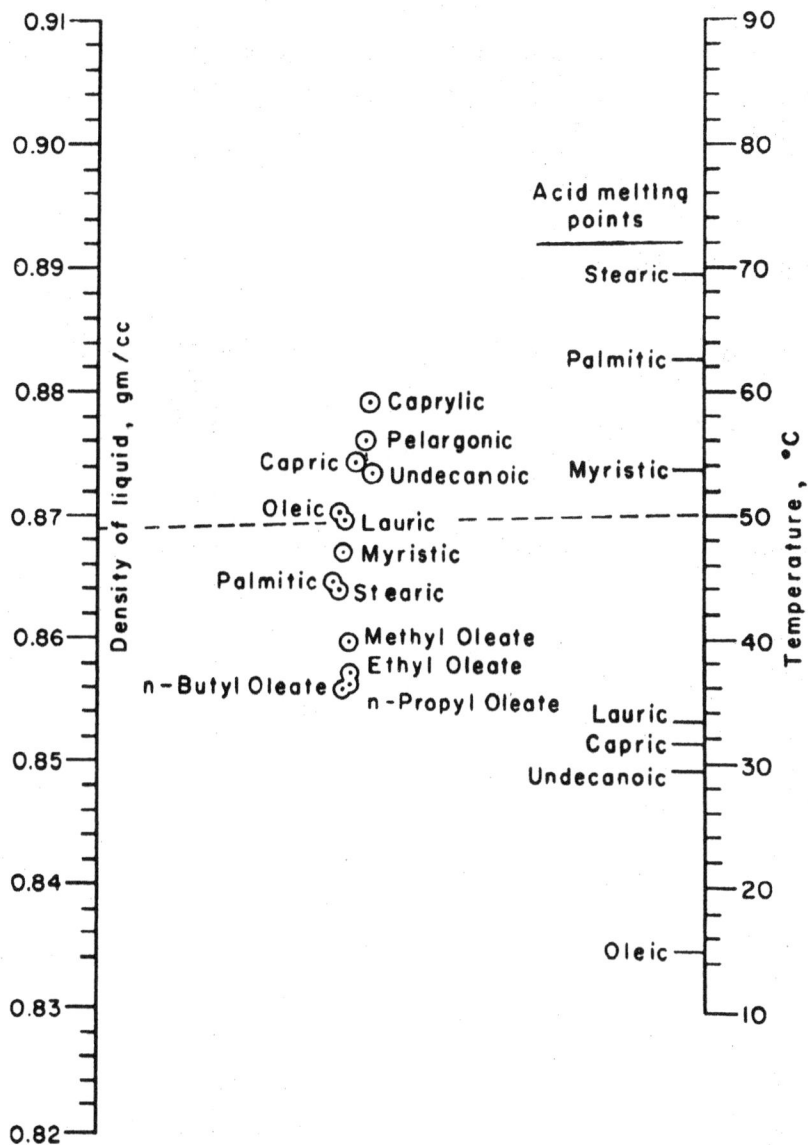

Figure 5-1

5-2   Specific Gravities of Commercial Formalin Solutions

GEORGE E. MAPSTONE

Figure 5-2 is based on published data[1] showing effects of formaldehyde and methanol contents on specific gravities of commercial formalin solutions.

Typical Example

The dashed line shows that a formalin solution containing 37.16% formaldehyde and 3.0% methanol has a specific gravity of 1.106 at 25°C; a solution containing 44.06% formaldehyde and 3.5% methanol has a specific gravity of 1.126 at 25°C.

[1]*Formalin Handling Manual*, Celanese Corporation of America, 1960.

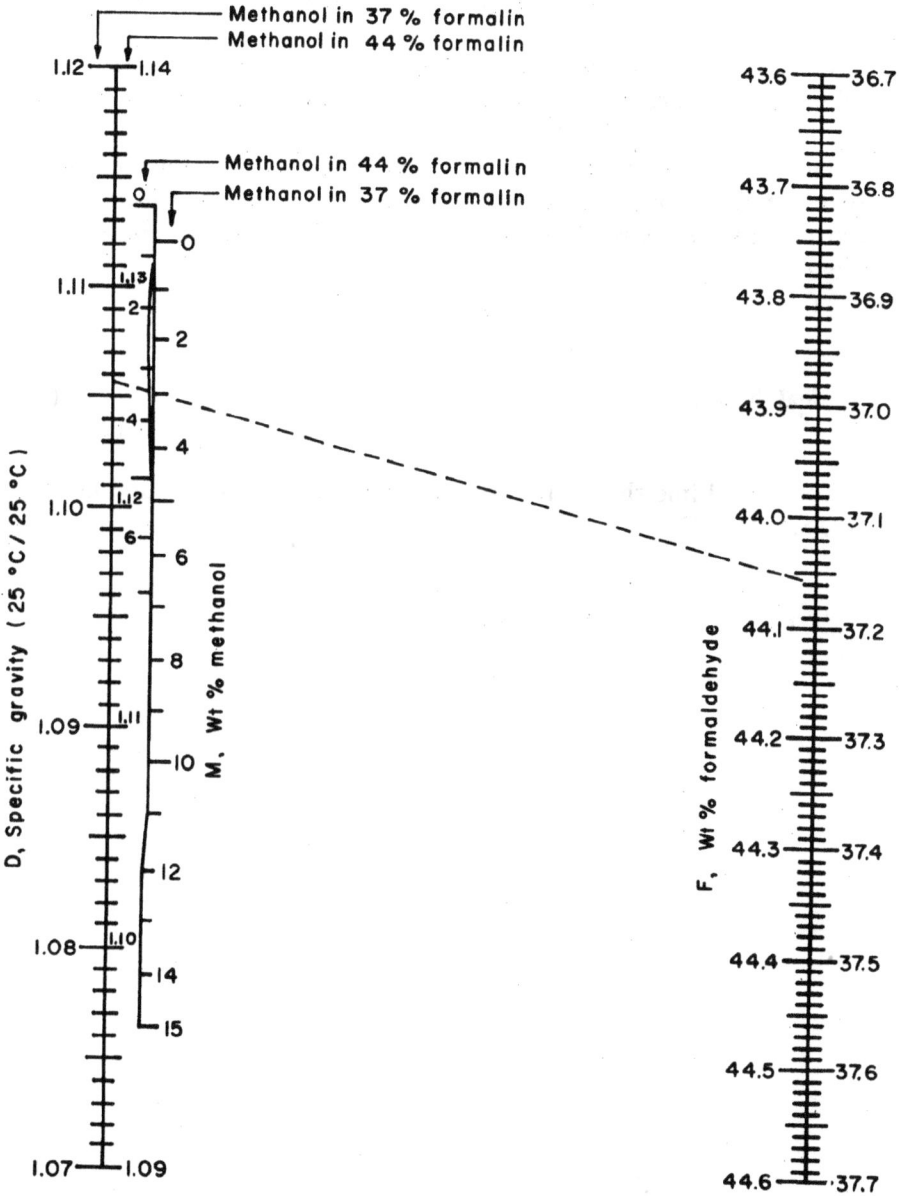

Figure 5-2

5-3 Specific Gravities of Ketones

GEORGE E. MAPSTONE

Published data[1] on the effect of temperature on the specific gravity of several ketones are presented in convenient nomographic form in Figure 5-3.

The chart gives the results with a specific-gravity accuracy of better than 0.001, except in the case of methyl ethyl ketone. In this instance, the maximum error is 0.002, because of the deviation of the relationship from linearity.

Typical Example

The dotted line shows that the specific gravity of ethyl butyl ketone is 0.835 at 2°C.

[1]*Ketones*, Union Carbide Corporation, 1964.

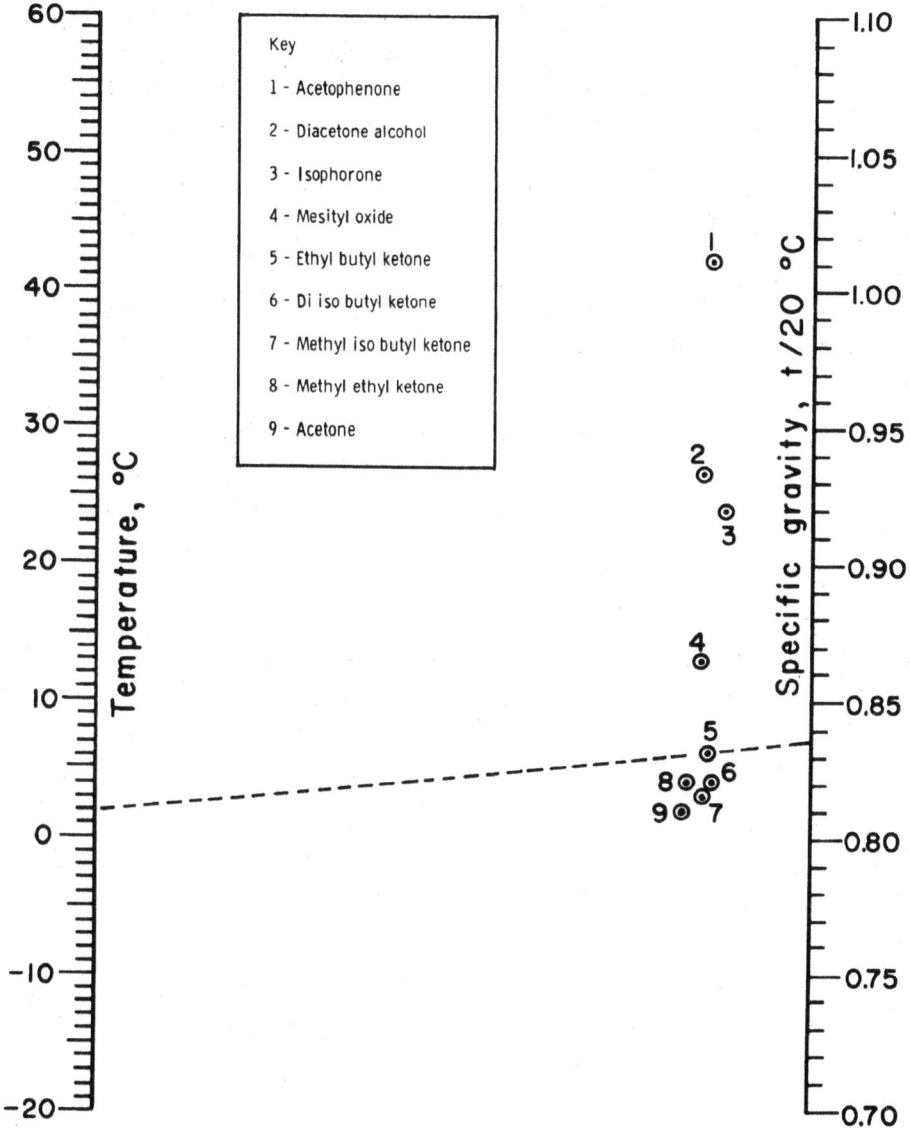

Figure 5-3

5-4   Specific Gravities of Acetone/Water Mixtures

GEORGE E. MAPSTONE

Published data[1] for the specific gravities of acetone-water mixtures are presented in Figure 5-4 in convenient nomographic form.   This chart and the original specific-gravity data agree within 0.002.

Typical Example

The dotted line shows that the specific gravity of a 60% solution (by wt.) of acetone in water is 0.909 at 10°C.

[1]*Ketones*, p. 7, Union Carbide Corporation, New York, 1964.

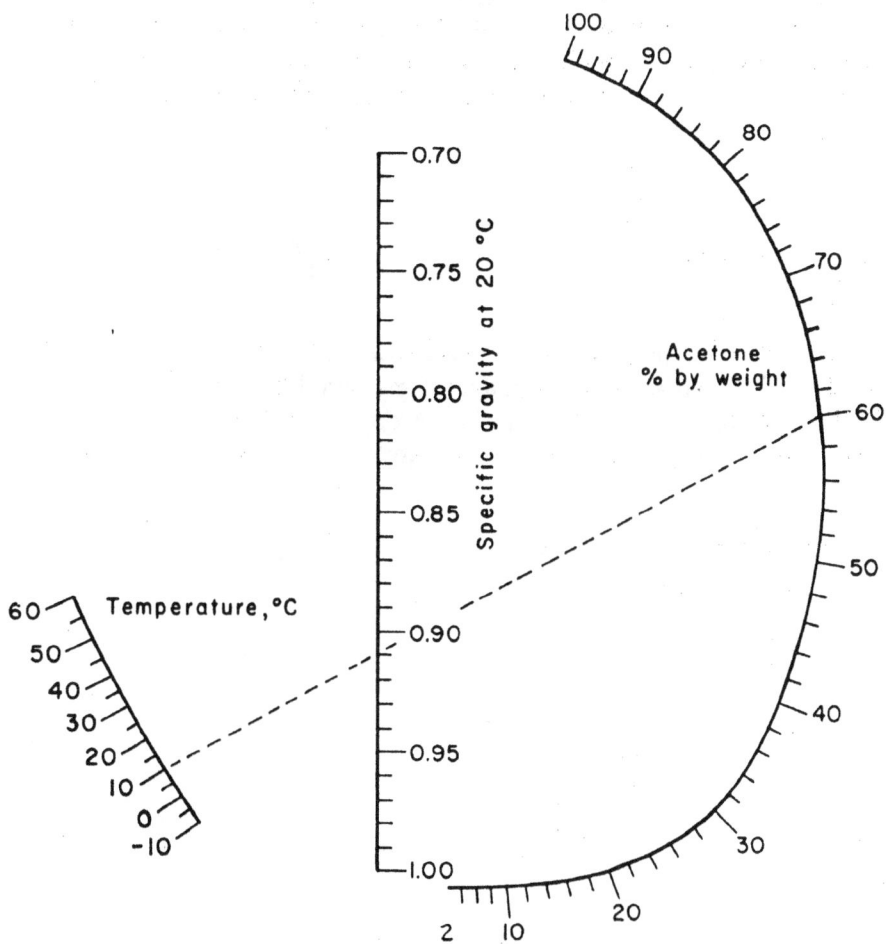

Figure 5-4

5-5   Density of Methanol-Water Solutions

STAFF, CHEMICAL PROCESSING

Figure 5-5 was prepared from data[1] on methanol-water solutions in the range of concentration from 25 to 100% methanol by weight. It provides a quick method of estimating methanol concentration, where density (or specific gravity) and temperature of the solution are known. The nomograph agrees closely (about 0.1%) with the tabulated data.

Typical Example

A sample of methanol-water solution has a density of 0.856 g/ml at 25°C. What is the percentage of methanol by weight?

The straight-edge connecting 0.856 on the density-scale with 25 on the temperature-scale, intersects to show 75% methanol in solution.

[1]*Synthetic Methanol*, Commercial Solvents Corp., New York.

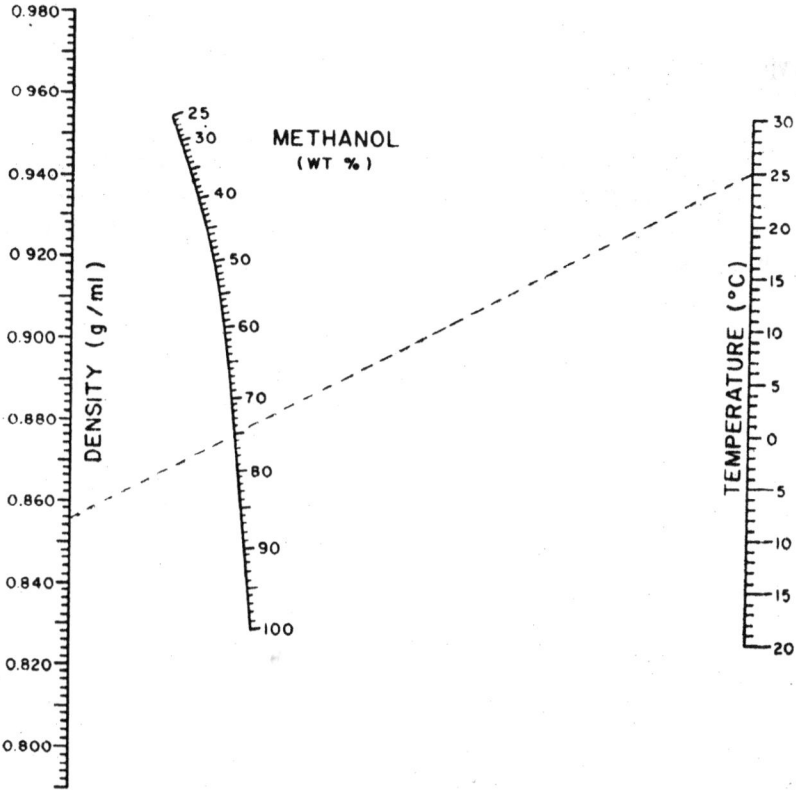

Figure 5-5

5-6   Temperature/Specific Gravity of Nitric
Acid Solutions

BILL SISSON

Figure 5-6 is used extensively by operators producing nitric acid for the fertilizer industry.

Typical Example ·

A straight line from 122°F (50°C) on temperature scale to 1.296 on the specific-gravity scale shows the HNO_3 concentration to be 54%. The chart can also be used to find specific gravity if percentage HNO_3 is known.

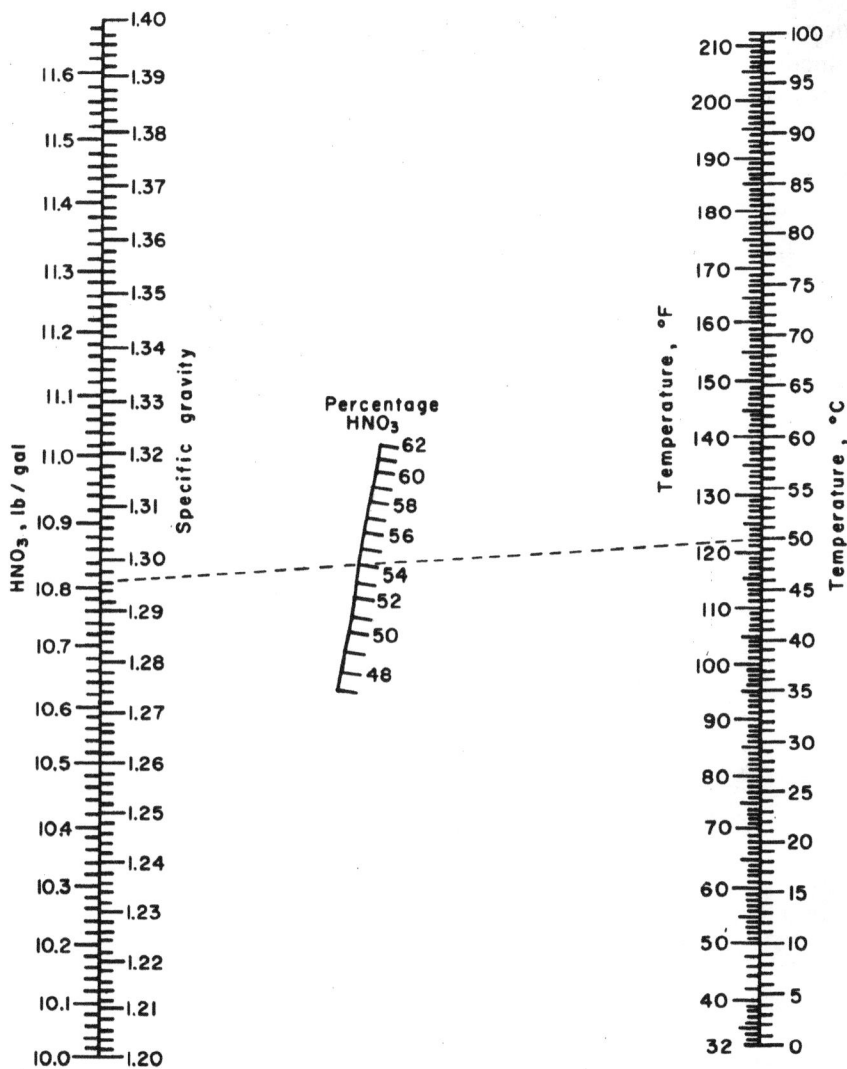

Figure 5-6

5-7   Specific Gravities of Aqueous Urea

BILL SISSON

Figure 5-7 is used hourly to check and control the urea concentration in the recovery unit of a large urea plant. Employed in conjunction with a chart showing the solubility of urea in water, it facilitates control of urea solution at desired temperature and concentration.

Typical Example

What is the urea concentration for a specific gravity of 1.172 and a temperature of 120°F? A straight line through these two points shows 65% (by wt.) urea.

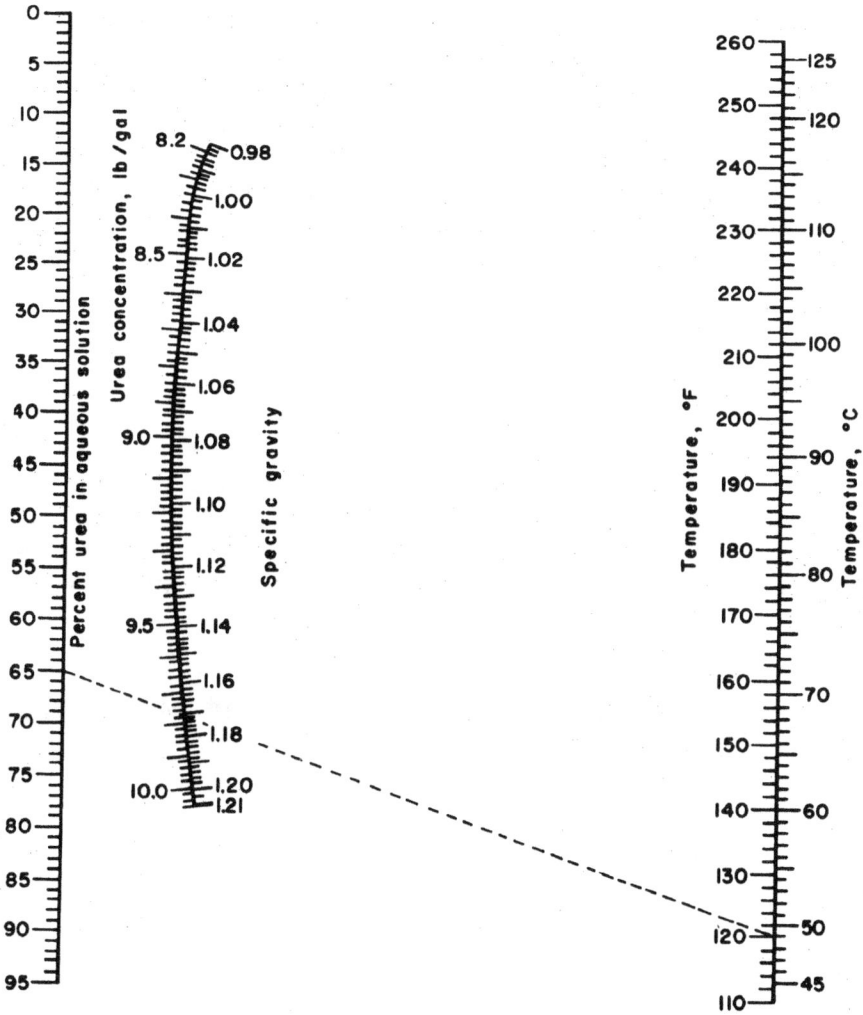

Figure 5-7

5-8   Alcohol Content of Ethanol-Water Mixtures from Specific Gravities

GEORGE E. MAPSTONE

It is often necessary to determine ethanol content of aqueous alcohol. Many tables are available which relate specific gravity under different conditions to ethanol content of mixture. However, the temperature coefficient of specific gravity is also a function of ethanol content.

Figure 5-8 has been drawn to allow ready estimation of alcohol content from observed specific gravities and temperatures on the basis of tabulated data for variations of both specific gravity at 60°F and the specific-gravity temperature coefficient with alcohol content.[1]

Typical Example

Use of the chart is illustrated by an example: An alcohol-water mixture has specific gravity of 0.881 at 85°F. What is the alcohol content? By means of a straight line connect 0.881 on the specific-gravity scale with 85 on the temperature scale. This line cuts the central alcohol content scale at 69% by volume.

[1]Pleeth, S. J. W., *Alcohol*, Appendices VII & VIII, pp. 248-249, Chapman & Hall Ltd., London, 1949.

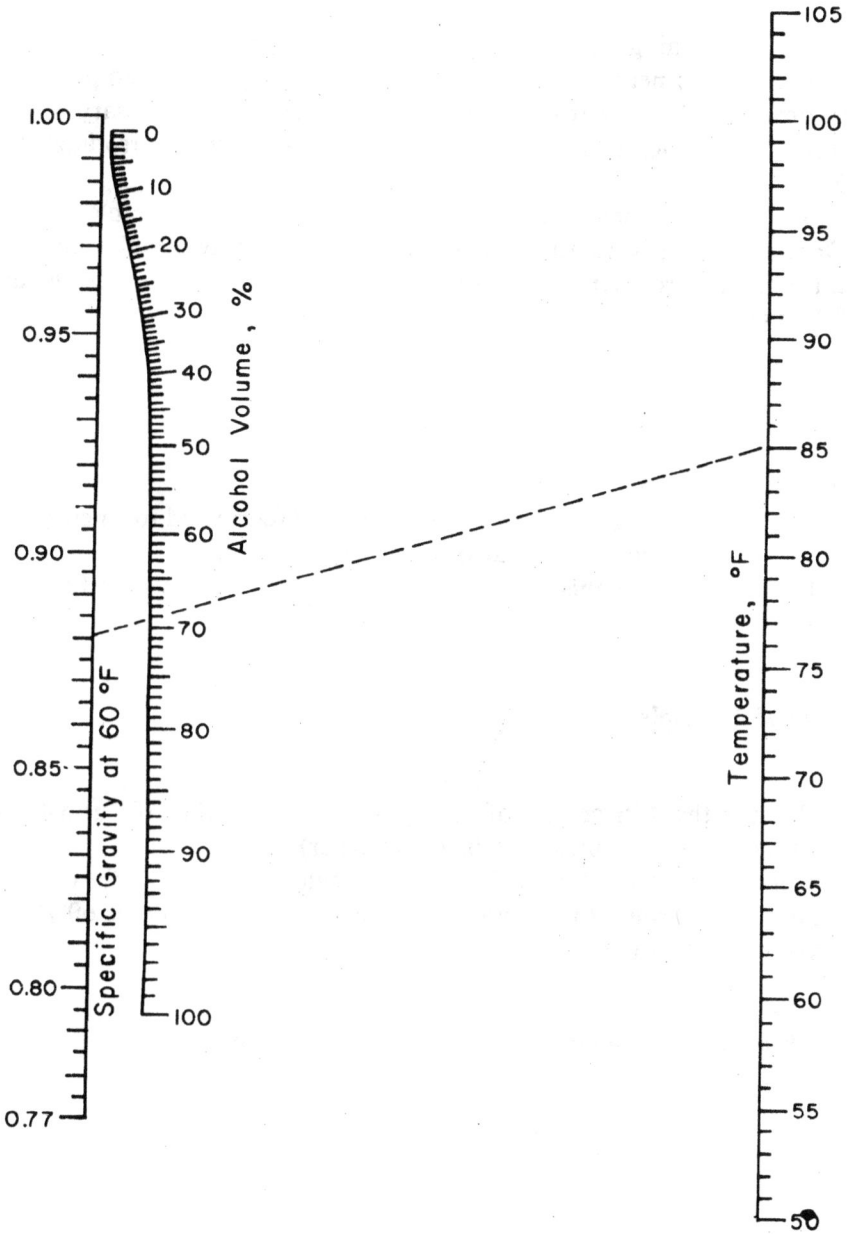

Figure 5-8

5-9   API Gravities of Asphaltic Materials

E. F. ROQUE

In determining API gravities of heavy hydrocarbon fractions (asphalt), it is necessary to have an extremely high bath temperature in order to obtain a reading on a hydrometer. It is necessary to go through a rather complicated procedure to correct the readings to 60°F.

In such cases, the manual[1] does not include a simplified table to use. Once a reading is obtained, it is necessary to read two tables and go through a three-step calculation. The calculation involved is as follows:

$$x = \frac{141.5}{\dfrac{y}{z}} - 131.5$$

where   x   = °API at 60°F

y   = sp. gr. at observed API at 60°F (corrected for meniscus)

z   = group factor at observed temperature

Figure 5-9 solves this equation in one setting of a straight edge.

Typical Example

What is the API gravity of an asphalt at 60°F if the API gravity is 10.0 (corrected for meniscus in hydrometer) at 300°F?

Connect 10.0 on the API observed scale (y) with 300°F on the temperature scale. Intersection with corrected API scale at 60°F (x) gives an API gravity of −1.5.

[1]*Fisher-Tag Manual for Inspectors of Petroleum*, 28th ed.

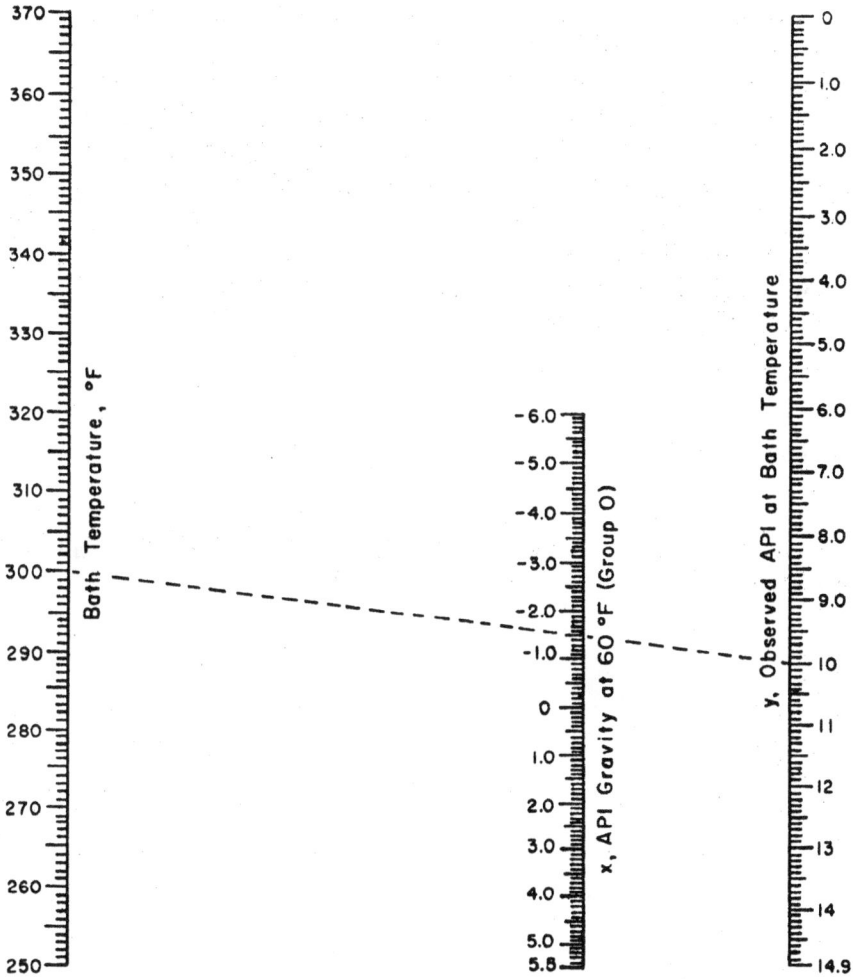

Figure 5-9

5-10   Specific Gravity of 25% Oleum

LESLIE L. VIZSALYI and HARRY A. ARBIT

In manufacture, storage and shipping of 25% oleum it is necessary to know the specific gravities of various concentrations of this material at different temperatures.

Through use of a standard reference (once sulfuric-acid content of the oleum is determined by titration), "percent 100% H_2SO_4" is converted to "% total SO_3" in one table. Next, "% total SO_3" is converted to specific gravity at a standard temperature in another table. Final conversion is to specific gravity at a desired temperature by applying correction factors.

Figure 5-10 gives directly the specific gravity of varying concentrations of 25% oleum at different temperatures.

Typical Example

At 86°F an oleum containing 105.84% sulfuric acid has a specific gravity of 1.921 (compared to water at 60°F).

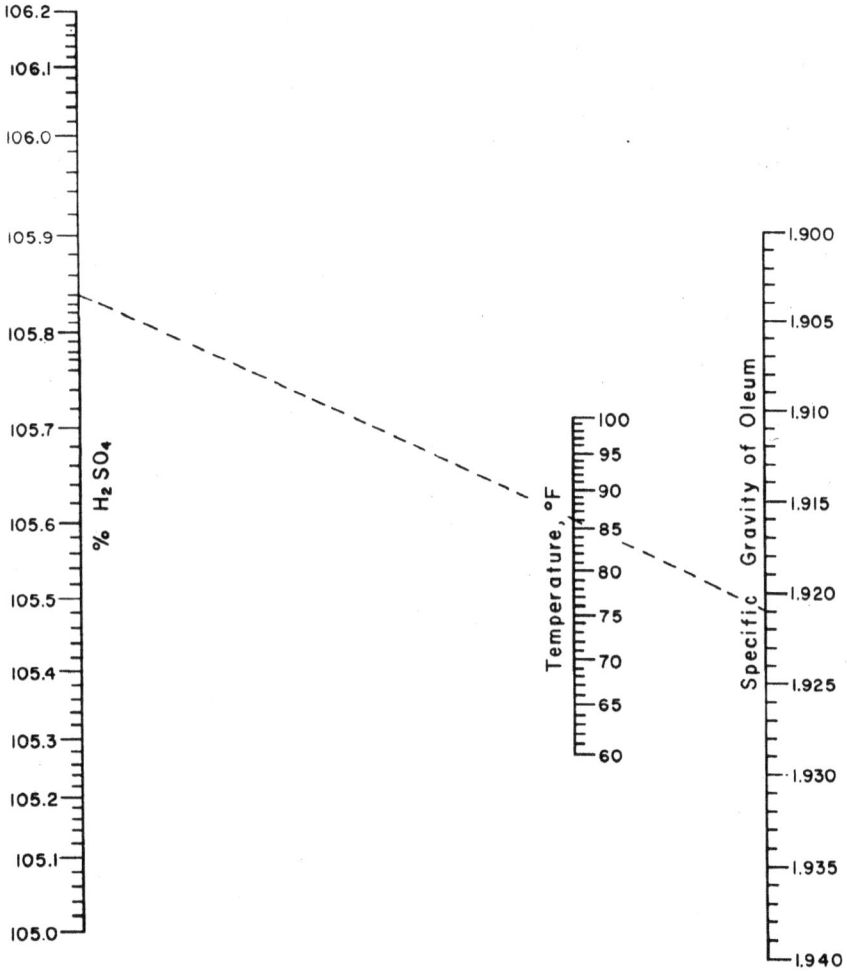

Figure 5-10

UNIT 6

Relationships of Pressure, Volume, and Temperature

Vapors at Low Pressures—Gases at Standard Conditions

Relationships of pressure, volume, and temperature are shown for vapors at low pressures, air, gases at standard conditions, and Freon-11. Wide but practical ranges of data characterize the charts.

6-1   Densities of Vapors at Low Pressures

FRED E. McKELVEY and D. S. DAVIS

Because 1 lb-mole of ideal gas at a pressure of 760 mm Hg and a temperature of 273° K occupies 359 ft³, the gas constant $R = PV/T = 760 (359)/273 = 999.5$, or approximately 1000. When the molecular weight is M lb, ϱ, the density of the gas in lb/ft³, equals PM/RT or $PM/1000T$, where P is the pressure in mm of mercury and T is the temperature in degrees Kelvin.

This equation can be solved readily and accurately by means of Figure 6-1. The chart is useful for specifying suitable vacuum-jet ejectors and for estimating pressure drops.

Typical Example

At a pressure of 110 mm Hg and a temperature of 500°K, what is the density of an ideal gas if the molecular weight is 90? Following the key, connect 90 on the M-scale and 110 on the P-scale with a straight line, marking the intersection with the α-axis. Connect this point and 500 on the T-scale with a straight line and read the density as approximately 0.0200 lb/ft³ on the ϱ-scale.

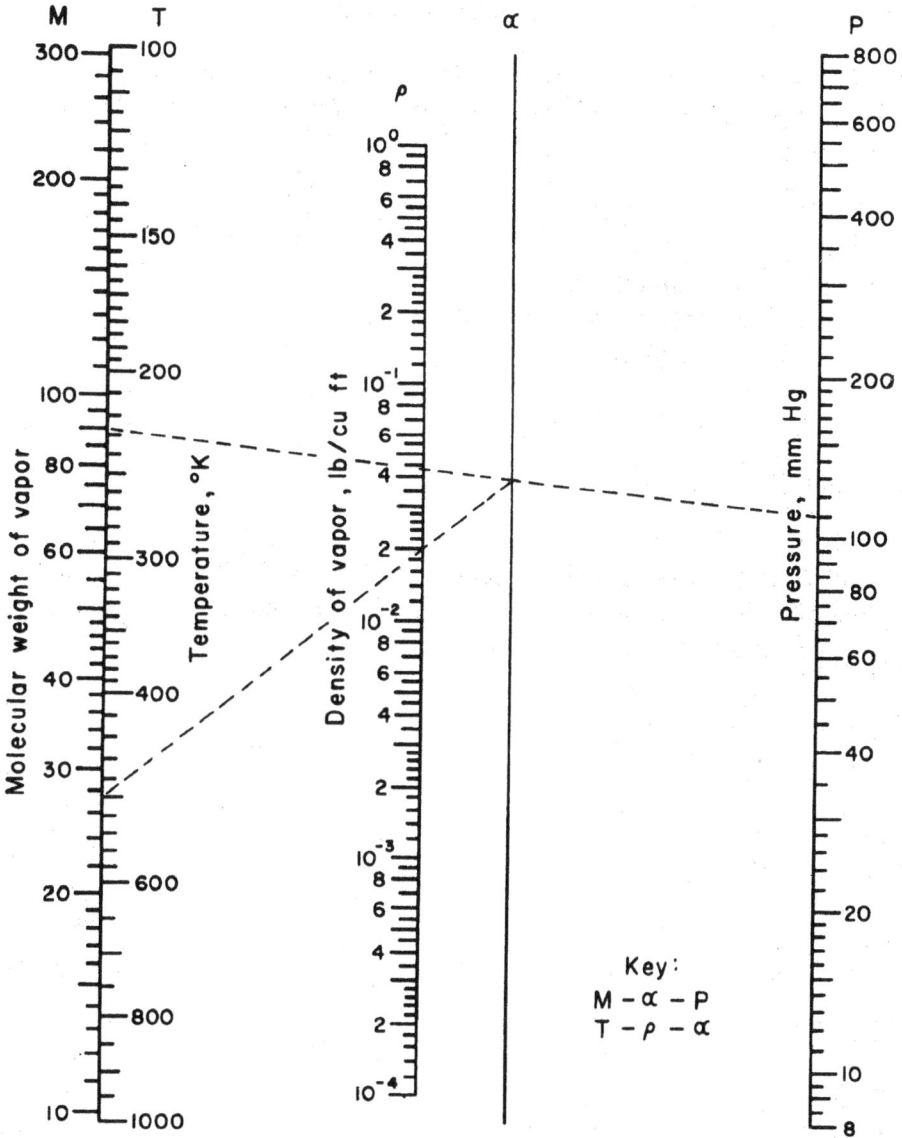

Figure 6-1

6-2   Determination of Specific Volume of Air at High Temperatures and Pressures

ELIZABETH SHROFF

Figure 6-2 is a convenient method for presentation and use of P-V-T data for air at high temperatures and pressures. It is based on the equation

$$0.01\,t + 2.795 = V^{1.25} \left[\frac{P}{61.5} \right]^{1.1875}$$

where   t   = temperature, °F

         V   = specific volume, ft^3/lb of air

         P   = pressure, lb/in.^{2} abs.

Typical Example

What is the specific volume of air at 300°F and 2500 lb/in.^{2} abs.? Join 300 on the t-scale with 2500 on the P-scale and read the value of 0.121 ft^3/lb of air on the V-scale.

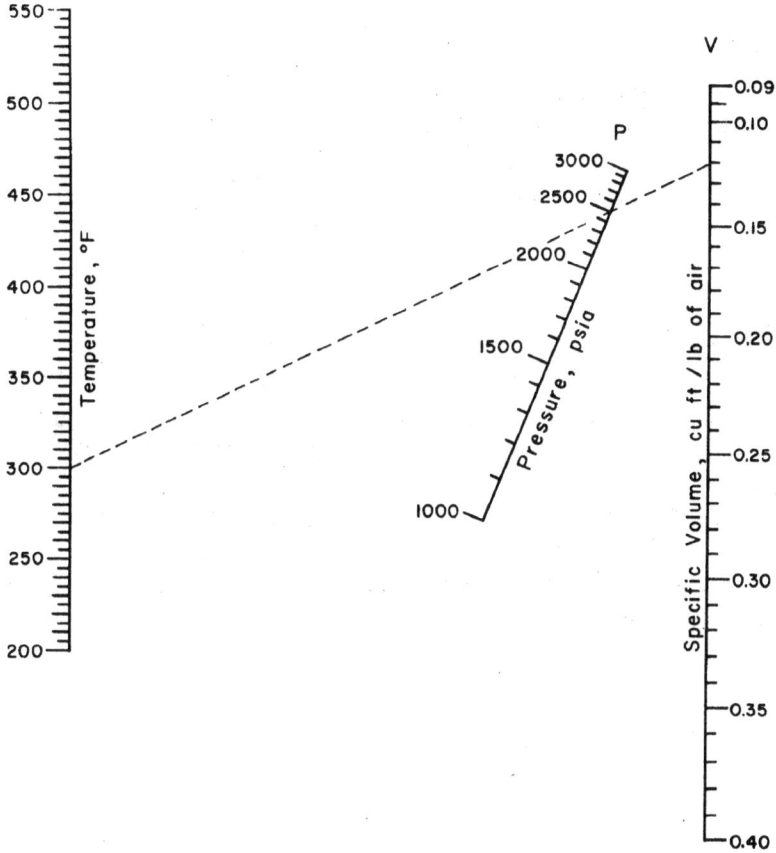

Figure 6-2

6-3 Adiabatic Expansion of Air

S. SALVA and L. RUSHING

Figure 6-3 permits a quick and simple determination of adiabatic expansion of air, eliminating calculations by gas-law formulas. It is based on the following equation:

$$P_1/P_2 = (V_2/V_1)^{1.41}$$

where P_1 = initial pressure, $lb/in.^2$
P_2 = final pressure, $lb/in.^2$
V_1 = initial volume, ft^3
V_2 = final volume, ft^3

Typical Example

Given an initial air pressure of 50 $lb/in.^2$ and an initial volume of 40 ft^3, what will be the final volume when a final pressure of 100 $lb/in.^2$ is applied?

Extend a line from 50 on the P_1-scale to 100 on the P_2-scale. The point at which this line intersects the reference line is now used as a pivot point. Draw a line from this point to 40 on the V_1-scale and read final volume as 24.5 ft^3 on the V_2-scale.

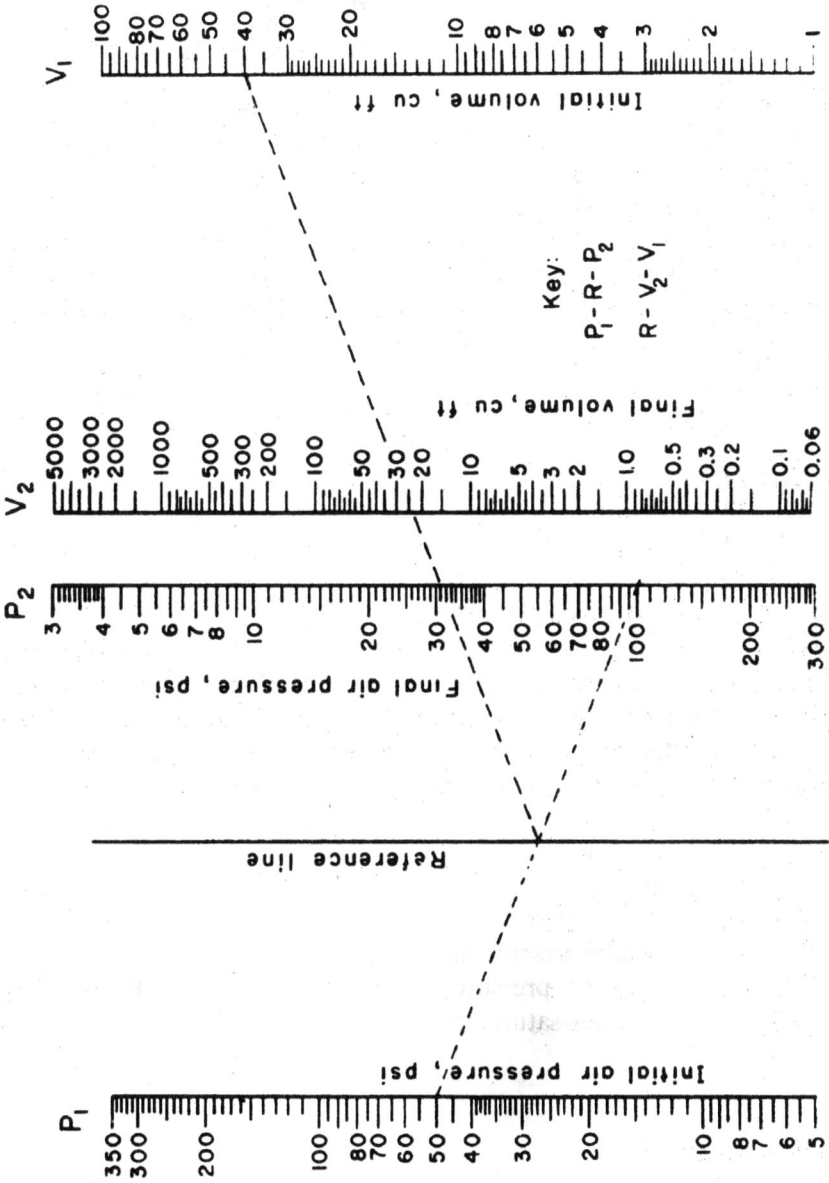

Figure 6-3

6-4   Correcting Gas Volumes to Standard Conditions

CHARLES R. NODDINGS and ROBERT M. LAWLESS

Although gas law calculations are simple, mistakes in algebraic sign and inversion of ratios can cause serious errors. Use of Figure 6-4 not only saves time in the calculation, but insures that all factors are considered and that inversions do not occur.

Based on the standard gas law, it can be shown that to reduce a metered gas flow to standard conditions (designated by subscript 2) from original (subscript 1) conditions, the following calculation is necessary:

$$V_2 = V_1 \left[\left[\frac{Z_2 R T_2}{P_2} \right] \left[\frac{P_1}{Z_1 R T_1} \right] \right] = V_1 \left[\left[\frac{P_1}{760} \right] \left[\frac{273}{T_1} \right] \left[\frac{1}{Z_1} \right] \right]$$

where P = pressure
          V = volume
          R = Boltzman's constant   } consistent units
          T = absolute temperature
          Z = compressibility factor

The terms inside brackets are a correction factor F, shown by this nomograph. If a wet test meter is used (or a dry meter on a wet gas), original pressure P_1 is composed of three parts: barometric and back pressure, and vapor pressure of water vapor contained. Therefore, this nomograph solves the equation

$$F = \left[\frac{P_{bar} + P_o - P_{vp}}{760} \right] \left[\frac{273}{273 + t} \right] \left[\frac{1}{Z} \right]$$

where   bar   = barometer
          o   = backpressure on system
          vp   = vapor pressure of water (gas temperature if gas is water-saturated)

Typical Example

The following data were gathered for a gas with a compressibility factor of 0.99, barometric pressure of 720 mm, back pressure of 30 mm, water vapor present to a partial pressure of 32 mm (saturated gas at 30°C), and temperature of 30°C. What factor will correct volumes measured under these conditions, to volumes at standard conditions?

Note that in the nomographic solution, of Figure 6-4, the baro-metric pressure of 720 mm is connected with the back pressure of 30 mm and the straight line extended to the reference line. The second straight line connecting this point with the water vapor pressure (or temperature for saturated gas) leads to a point on the second reference line. Following the transition guide lines, move the point to the adjoining reference line.

When a straight line has been extended from this point through the proper compressibility factor, the final reference point is known. The line connecting this point to the gas temperature-scale crosses the F-scale at 0.86. The factor of 0.86, then, is the correction factor by which the observed volume must be multiplied, to produce the actual volume at standard conditions of 0°C and 760 mm Hg.

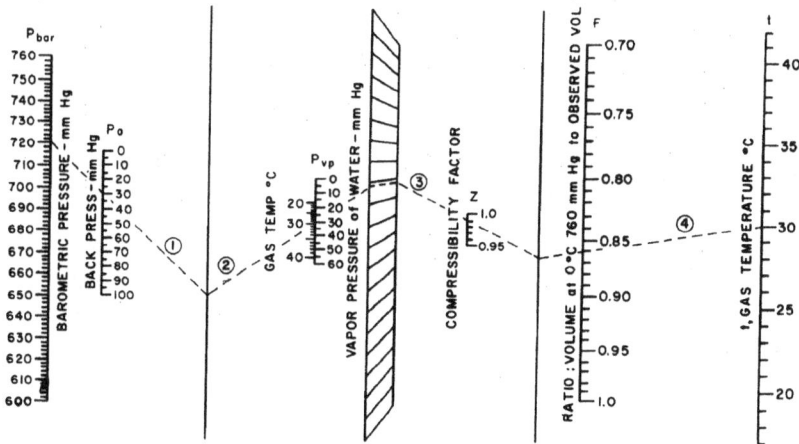

Figure 6-4

6-5   Pressure-Volume-Temperature Relationships
for Freon-11

ROBERT L. BREHM

Pressure-volume-temperature relationships for superheated Freon-11 (trichloromonofluoromethane)[1] can be expressed by the equation

$$V = \frac{t + 460 - 2.066\,P + 0.00392\,P^2}{13.84\,P^{0.9592} - 1.0}$$

where   V = specific volume ft^3/lb
        t = temperature °F
       P = pressure, lb/m^2

Figure 6-5 provides rapid, accurate solution of the equation.

Typical Example

What is the specific volume of Freon-11 at 300°F and a pressure of 1.2 lb/in.^{2}?   Connect 300 on the t-scale with 1.2 on the P-scale.   Read the specific volume as 49.0 ft^3/lb on the V-scale.

[1]Perry, J. H., *Chemical Engineers' Handbook*, 3rd ed., McGraw-Hill Book Co., New York, 1950.

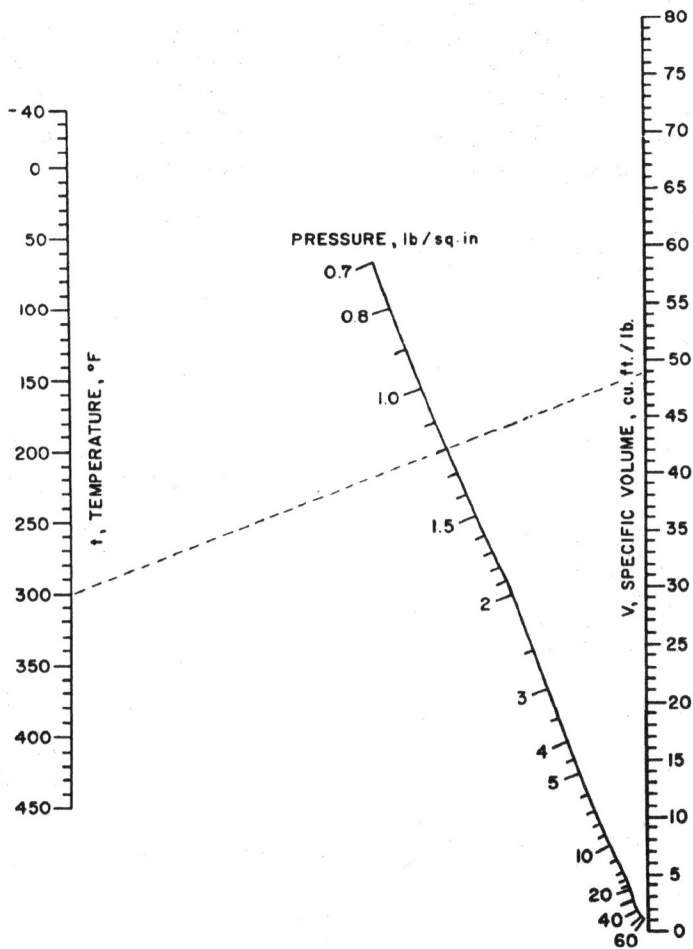

Figure 6-5

6-6   Pound-Molecular Volume of Gases

GERALD TEPLITZKY

Figures 6-6a and 6-6b enable one to calculate the pound-molecular volume of ideal gases in ft³ when the temperature in °C and the pressure in mm Hg are known.

The index line in Figure 6-6a shows that the pound-molecular volume of an ideal gas is approximately 23,000 ft³ when the temperature is 300°C and the pressure is 25 mm Hg.

The index line in Figure 6-6b shows that the pound-molecular volume of an ideal gas is approximately 2500 ft³ at 250°C and 210 mm Hg.

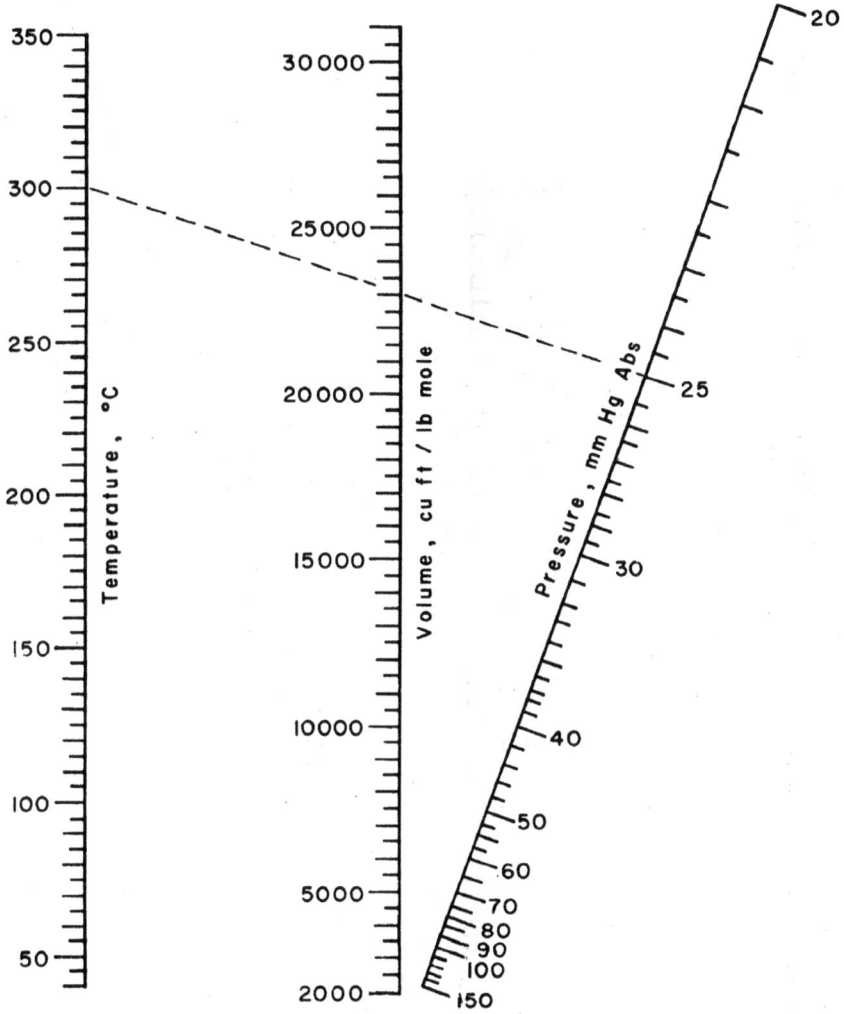

Figure 6-6a

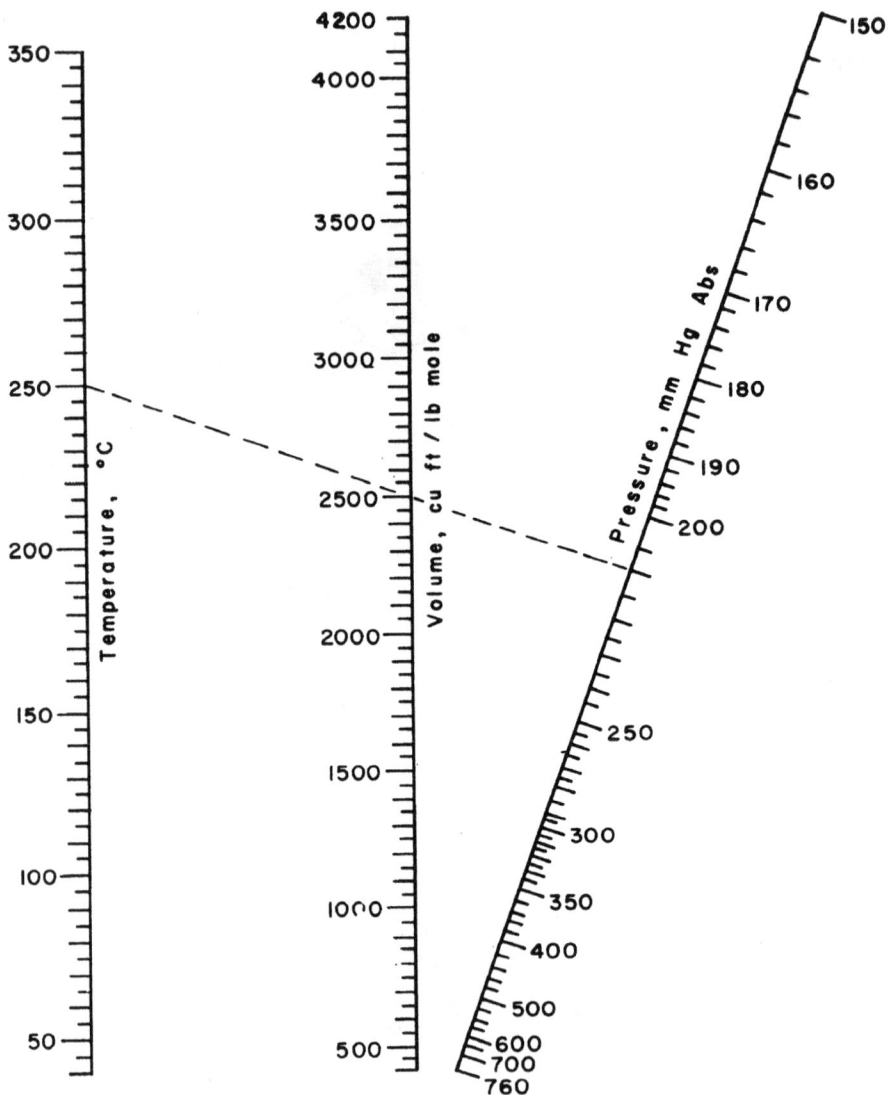

Figure 6-6b

UNIT 7

Sizing of Equipment

Cyclone Dip Legs—Catalyst Hoppers—Pumps and Pump Motors

These six nomographs help the engineer in choosing the proper sizes for cyclone dip legs, catalyst hoppers, pumps, and pump motors to meet industrial requirements. One chart enables him to determine the discharge velocity of a centrifugal pump.

7-1   Sizing of Cyclone Dip Legs

ELIZABETH SHROFF

Figure 7-1 provides convenient means for determining primary cyclone dip leg diameter in reactors and regenerators of fluid catalytic cracking units. Sizes of secondary and tertiary cyclone dip legs are normally not critical and are therefore usually fixed.

The chart is based on the equation

$$\frac{n}{144} \cdot \frac{\pi D^2}{4} = \frac{3600 \, Vd}{G}$$

For normal design catalyst mass velocity G of 270,000 lb/(hr)(ft²), the equation rearranges to

$$D = \left[\frac{7.68 \, Vd}{\pi \, n} \right]^{1/2}$$

where   D   =   diameter of dip legs, in.
            V   =   vapor velocity, actual ft³/sec
            d   =   dilute phase density, lb/ft³
            n   =   number of primary cyclones

Typical Example

Size the primary dip legs of the cyclones in a regenerator with 18 primary stages, handling 2700 actual ft³/sec of vapors with a catalyst entrainment density of 0.5 lb/ft³.

Connect 2700 on the V-scale with 18 on the n-scale to intersect the reference line. Join 0.5 on the d-scale with the reference line intersection point and read on the D-scale a dip leg diameter between 13 and 14 in., therefore, select 14 in. standard pipe for the primary dip legs.

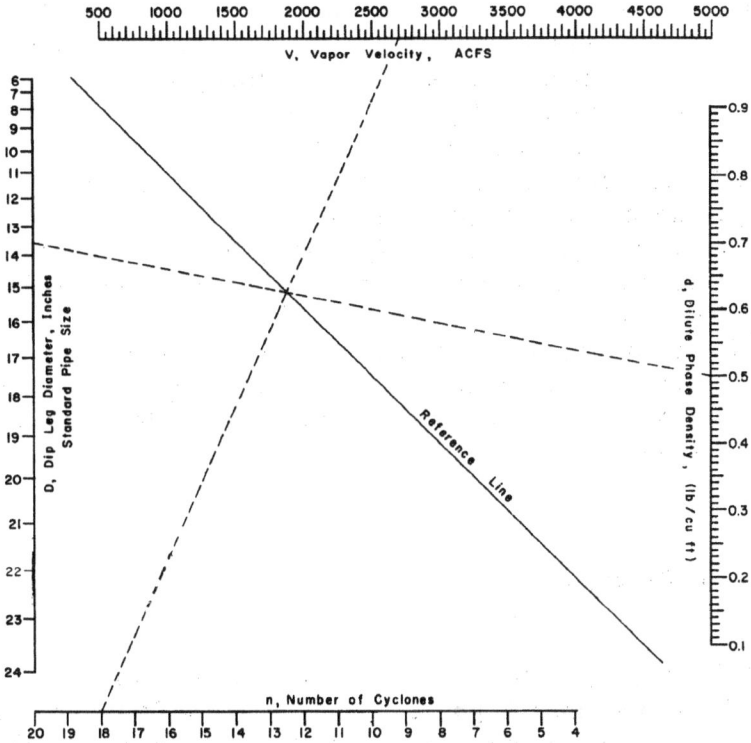

Figure 7-1

7-2   Nomograph for Sizing Catalyst Hoppers

ELIZABETH SHROFF

An aid in the sizing of catalyst hoppers, Figure 7-2 is based on the equation

$$T = \frac{\pi D^2 \rho}{8000} \left(L + \frac{D}{6} \cot a\right)$$

where   T = capacity of hopper, tons
        D = diameter of hopper, ft
        L = length of hopper (tang. to tang.), ft
        a = angle of sloping sides of conical bottom (30°)
        ρ = density of catalyst, lb/ft^3

The formula does not include the volume in the spherical head, as this is available for outage space.

This nomograph essentially solves two types of problems:

1) To size a hopper that is to hold a volume of catalyst of known density.

2) To find the volume of catalyst of known density that can be stored in an existing hopper whose length and diameter are known. (Key given with nomograph is for this type of problem.)

Typical Examples

Index line (1) shows that 755 tons of catalyst with density of 40 lb/ft^3 would require a hopper 25 ft in diameter and 70 ft long. It may be seen that various solutions could be found. For instance, another alternate would be a hopper 27 ft in diameter 58.5 ft long.

Index line (2) shows that capacity of a hopper 40 ft long and 15 ft in diameter is 117.5 tons for a 30 lb/ft^3 density catalyst.

Note: Use ρ_1-scale with T_1-scale and ρ_2-scale with T_2-scale.

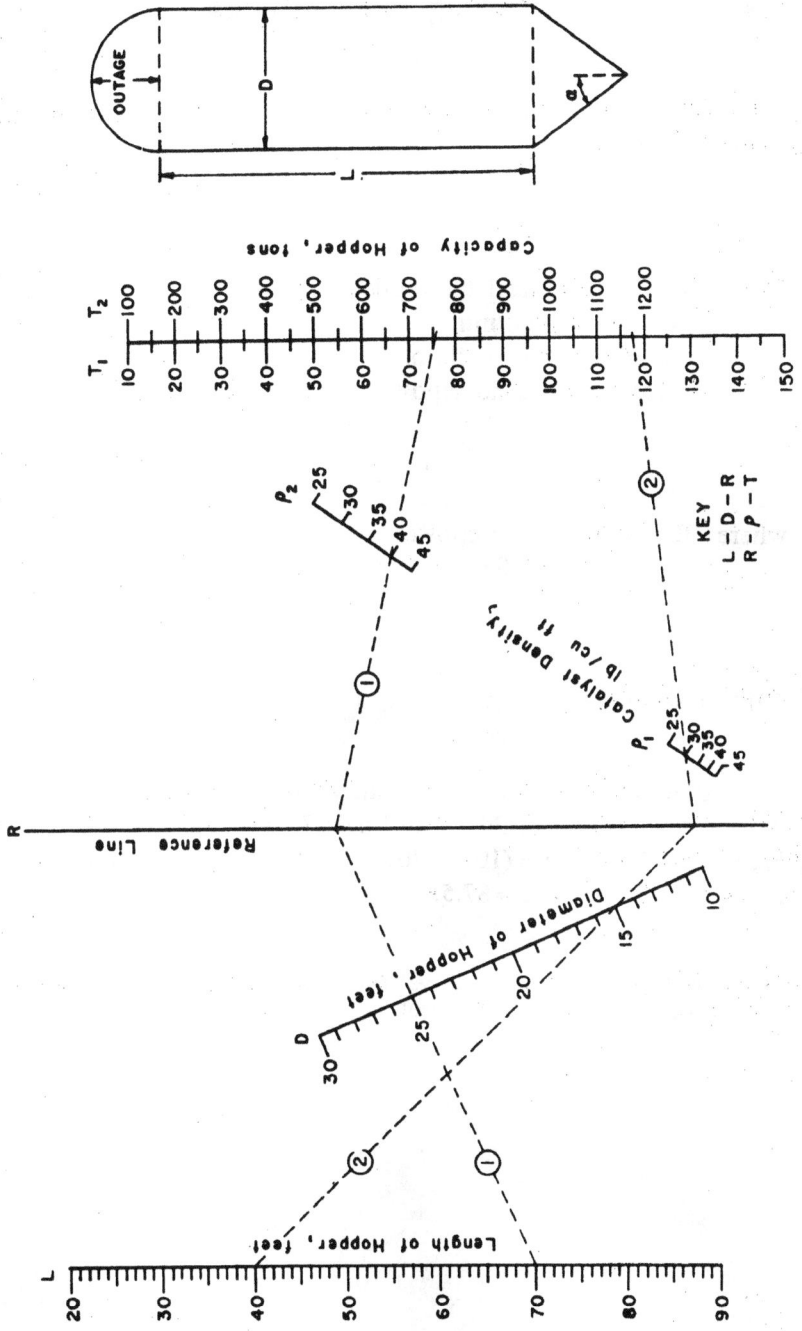

Figure 7-2

7-3   Sizing Pump Motors

F. CAPLAN

The theoretical hydraulic horsepower of a 100%-efficient pump is given by the equation[1]:

$$T = \frac{GH_p}{1714}$$

where T = theoretical hydraulic horsepower
G = flow, gal/min
H_p = total head, lb/in.^{2}

The actual or brake horsepower is given by the equation

$$B = \frac{T}{E} \times 100$$

where B = brake horsepower
E = pump efficiency, %

Typical Example

What brake horsepower is required for an 80%-efficient pump to deliver 500 gal/min at 240 lb/in.^{2} head? Align $G = 500 = 10(50)$ with $H_p = 240$ and read $T = (10)7 = 70$; align T with $E = 80$ and read brake horsepower $= (10)\ 8.75 = 87.5$.

[1]Perry, J. H., *Chemical Engineers' Handbook*, 3rd ed. pp. 1416-17, McGraw-Hill Book Co., New York (1950).

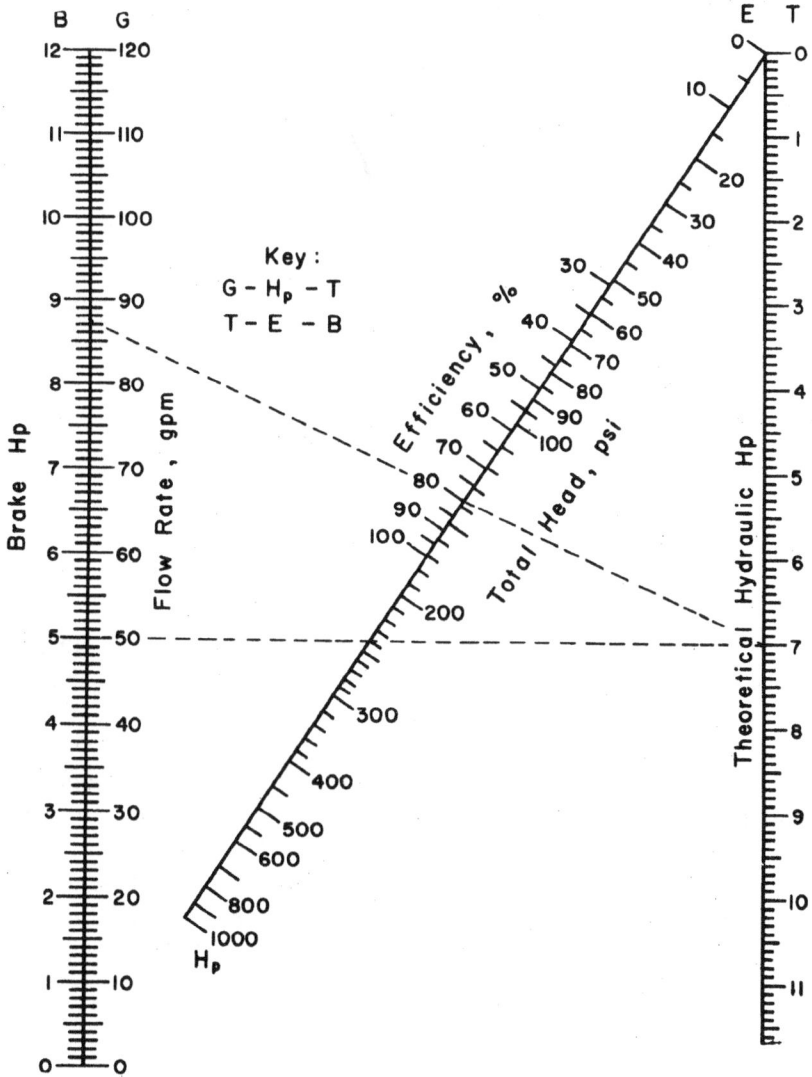

Figure 7-3

7-4   Sizing of Pumps

DAVID W. KUHN

Figure 7-4 for sizing pumps quickly is based on the equations

$$\Delta H = 2.31 \frac{\Delta P}{s} \tag{1}$$

$$THP = \frac{8.33 \, Rs \, \Delta H}{33,000} = 5.83 \, (10^{-4}) \, R\Delta P \tag{2}$$

$$THP = \frac{\eta \, BHP}{100} \tag{3}$$

where ΔH = discharge head, ft of fluid
        ΔP =   pressure increase across pump, lb/in.^{2}
         s   =   specific gravity of fluid
        R   =   rate of discharge, gal/min
        η =   pump efficiency, %
THP and BHP are theoretical and brake horsepowers, respectively.

Typical Example

Hexane (specific gravity, 0.66) is pumped at 700 gal/min, with a discharge head of 100 ft and a pump efficiency of 48%. Find the pressure increase and the theoretical and brake horsepowers.

For convenience, multiply the head by 10, obtaining 1000. Connect 1000 on the ΔH-scale and 0.66 on the s-scale with a straight line. Read the pressure increase as 286 lb/in.^{2} on the ΔP-scale. Connect this point and 700 gal/min on the R-scale with a straight line. Read 117 as the theoretical horsepower on the THP-scale. Connect this point and 48 on the η-scale with a straight line, and find the brake horsepower to be 243 on the BHP-scale. Divide 243 by 10 to obtain 24.3 as the brake horsepower for 100 ft of head.

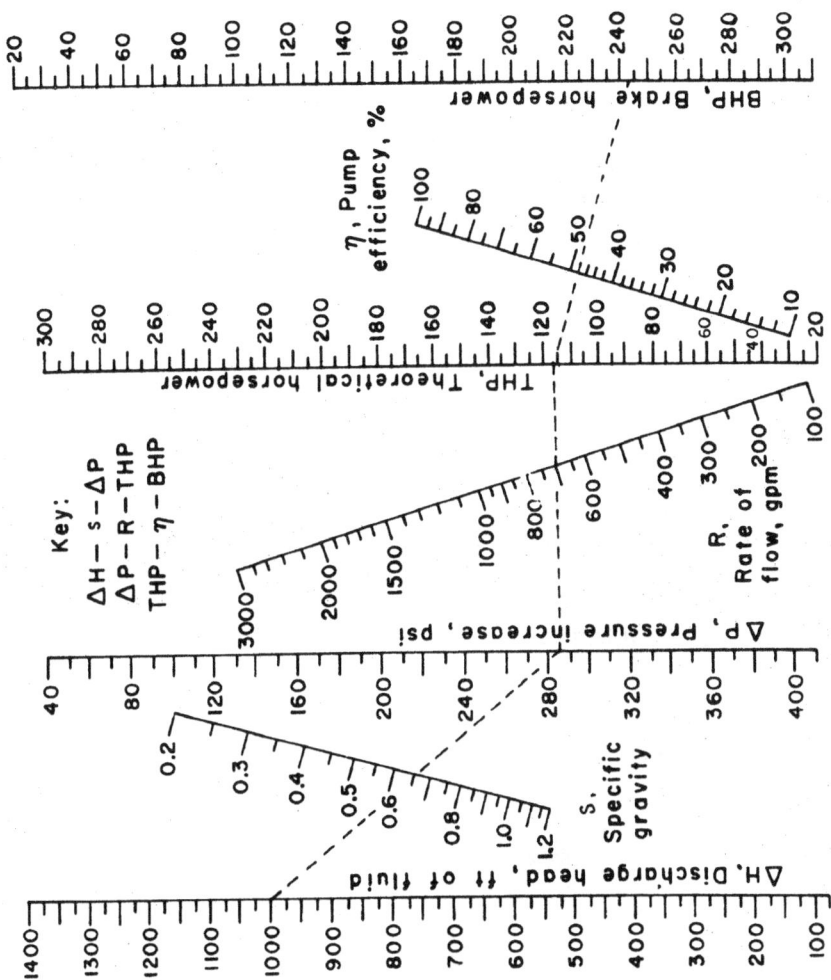

Figure 7-4

7-5   Theoretical Pump Horsepower

S. E. HENRY

Figure 7-5 provides a graphical solution for the following equation:

$$P = VHS/3960$$

where P = theoretical pump horsepower
        V = volume of liquid, gal/min
        H = head, ft of liquid
        S = specific gravity of liquid

Typical Example

$$V = 5000 \text{ gal/min}$$
$$S = 0.95$$
$$H = 60 \text{ ft}$$

Connect 0.95 on the S-scale and 60 on the H-scale with a straight line. Note the point where this line intersects the T-axis. Extend a straight line from 5000 on the V-scale through the point just found until the line intersects the P-scale at 72 horsepower.

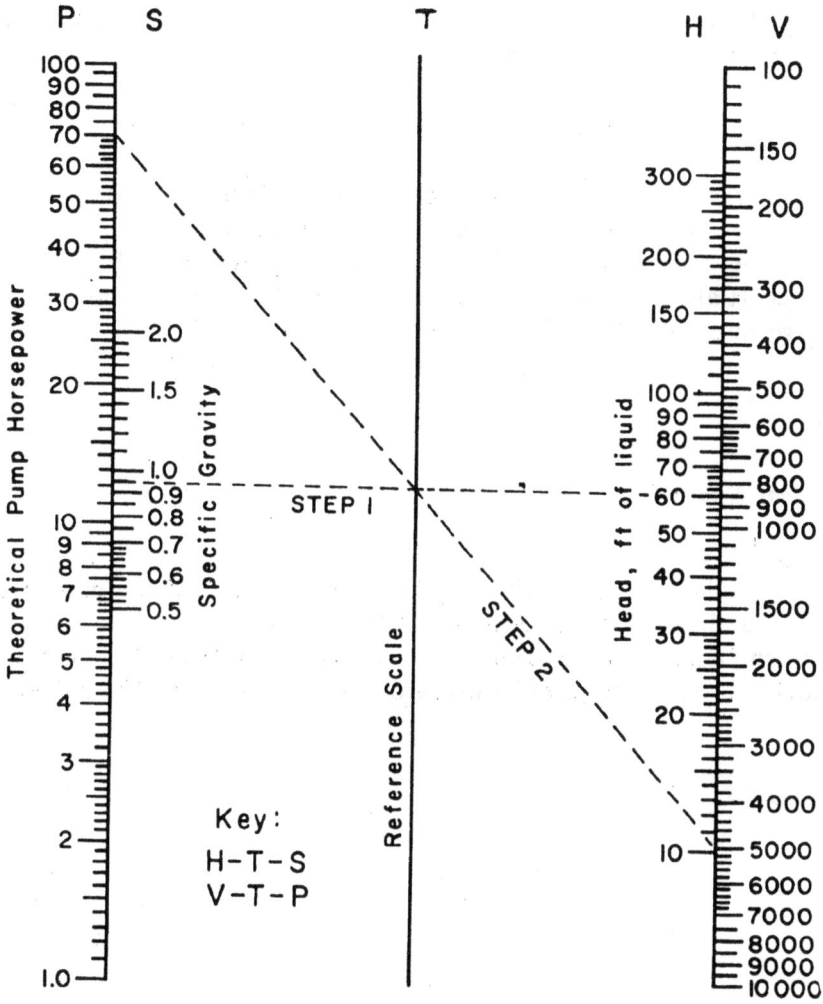

P   S   T   H   V

Theoretical Pump Horsepower

100
90
80
70
60
50
40
30
2.0
20
1.5
Specific Gravity
1.0
0.9
0.8
0.7
0.6
0.5
10
9
8
7
6
5
4
3

Key:
H-T-S
V-T-P

2

1.0

STEP I

Reference Scale

STEP 2

Head, ft of liquid

100

300
200
150
100
90
80
70
60
50
40
30
20
10

100
150
200
300
400
500
600
700
800
900
1000
1500
2000
3000
4000
5000
6000
7000
8000
9000
10 000

Figure 7-5

7-6   Discharge Velocities of Centrifugal Pumps

S. J. SALVA and L. J. CRULL

Figure 7-6 permits quick determination of discharge velocities of centrifugal pumps. It is based on the equation

$$V = \frac{0.321\ Q}{A}$$

where V = velocity of discharge of a centrifugal pump, ft/sec
       Q = quantity of water discharged, gal/min
       A = cross-sectional area of discharge nozzle, in.^{3}
The chart also has a scale for converting cross-sectional areas of discharge nozzles to their diameters.

Typical Example

What will be the exit velocity if the area of the discharge nozzle is 10 in.^{2} and the water flow is 400 gal/min?
Extend a straight line from 400 on the Q-scale to 10 on the A-scale. Read the velocity where this line crosses the V-scale as 12.8 ft/sec.

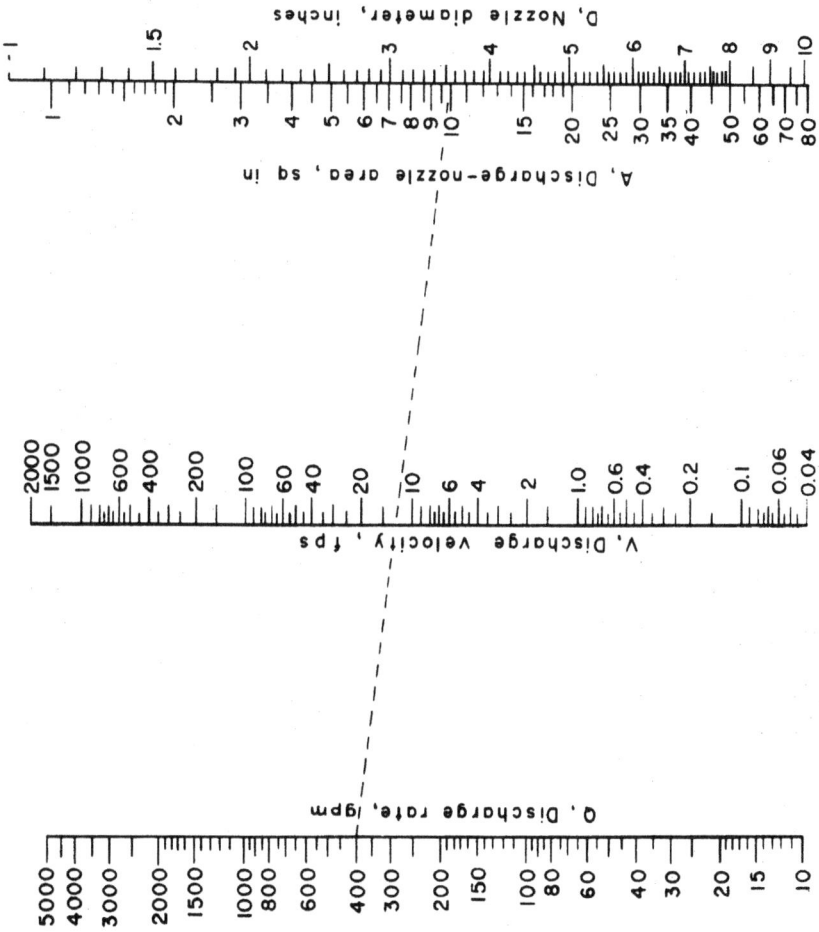

D, Nozzle diameter, inches

A, Discharge-nozzle area, sq in

V, Discharge velocity, fps

Q, Discharge rate, igpm

Figure 7-6

UNIT 8

Miscellaneous

Properties—Equipment—Conversion Factors

This unit consists of nomographs that deal with solubilities, viscosities, factors of various kinds, blending, industrial production, crystallization, electrodeposition, contaminants, costs, design of vessels, purging, screening, sedimentation, relative volatility, efficiencies, permeabilities, specimen geometry, determination of water in many materials, air for air lifts, bag filters, blowdown of boilers, compressors, and many others.

8-1   Vapor Composition in Ammonia Condensers

GEORGE E. MAPSTONE

If an inert or non-condensible gas is present in anhydrous ammonia vapor being passed to a condenser, some ammonia will remain in the vapor phase. This is because the inert constituent becomes saturated with ammonia from the vapor pressure of the condensed liquid. The gaseous phase must then be vented in order to maintain proper operation of the condensing system.

The quantity of ammonia thus lost from the system is a function of the amount of inert gas present, the pressure of the system, and the vapor pressure of the condensed liquid ammonia. This relationship is represented by the equation[1]

$$R = \frac{p}{P-p}$$

where R = ratio of ammonia vapor to inert gas in the purge
P = pressure of the system, lb/in.^{2}
p = vapor pressure of ammonia at the condenser temperature, lb/in.^{2}

The equation can be revised to the form

$$\frac{p}{P} = \frac{R}{R+1}$$

This relationship is presented in Figure 8-1. From published vapor-pressure data for ammonia, the vapor-pressure scale has also been scaled as receiver temperature.

Typical Example

What is the ratio of ammonia vapor retained per volume of inert gas present if the receiver temperature is 35°C and total pressure of the system is 260 lb/in.^{2}? The broken index line shows that 3 volumes of ammonia vapor will be present for each volume of inert gas.

[1]Ross, T. K., and Freshwater, D. C., *Chemical Engineering Data Book*, London, 1958, p. 334.

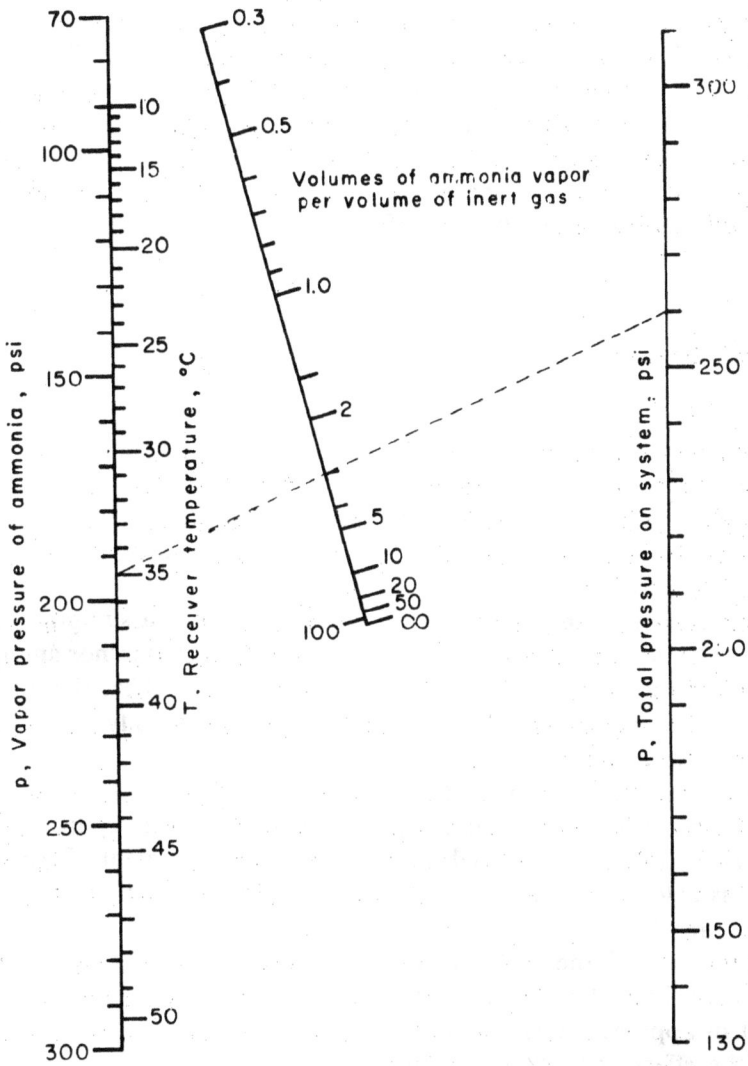

Figure 8-1

8-2   Aromatics Content of Benzols
and Naphthas

George E. MAPSTONE

Benzols and aromatic naphthas, whether obtained from coal tars or from petroleum sources, vary appreciably in composition depending on production method and source material. Figure 8-2 has been prepared to give a quick and reasonably accurate estimate of aromatics content of the sample from the specific gravity at 15.5°C and the mean boiling point. This can be conveniently taken as the 50% point of standard distillation at 760 mm Hg.

Typical Example

What is the aromatics content of a benzol which distills 50% at 87.5°C and which has a specific gravity of 0.847? Connect 87.5 on the boiling-point scale with 0.847 on the specific-gravity scale by means of straight edge. Extend to cut bottom scale at 83% aromatics (*i.e.*, 17% paraffins).

Limitations: The chart has been designed on the assumption that

1) The presence of olefins, napthenes, and the usual minor amounts of nonhydrocarbons (e.g. thiophene, etc.) does not affect the result.

2) Different classes of nonaromatics are uniformly distributed throughout the boiling range.

With respect to the first assumption, an examination of the limitations of the chart when applied to mixtures of pure hydrocarbons showed that olefins increased the effective paraffin content of the mixture (as calculated from the H:C ratio), whereas naphthenes had the reverse result.

Magnitudes of these effects were about equal and of the order of 8% of the olefin and naphthene contents (i.e. 10.0% of olefins had the effect of approximately 10.8% of paraffins and 10.0% of naphthenes had the effect of 9.2% of paraffins).

In the presence of both classes of materials, these effects tend to cancel out, depending on their relative proportions. Although olefins and naphthenes seldom occur in equal proportions in benzols and naphthas, errors attributed to this factor seldom reach 2%. This is satisfactory for most purposes.

The second assumption—that non-aromatics are uniformly distributed throughout the boiling range—has generally been found to be true within the precision of the methods.

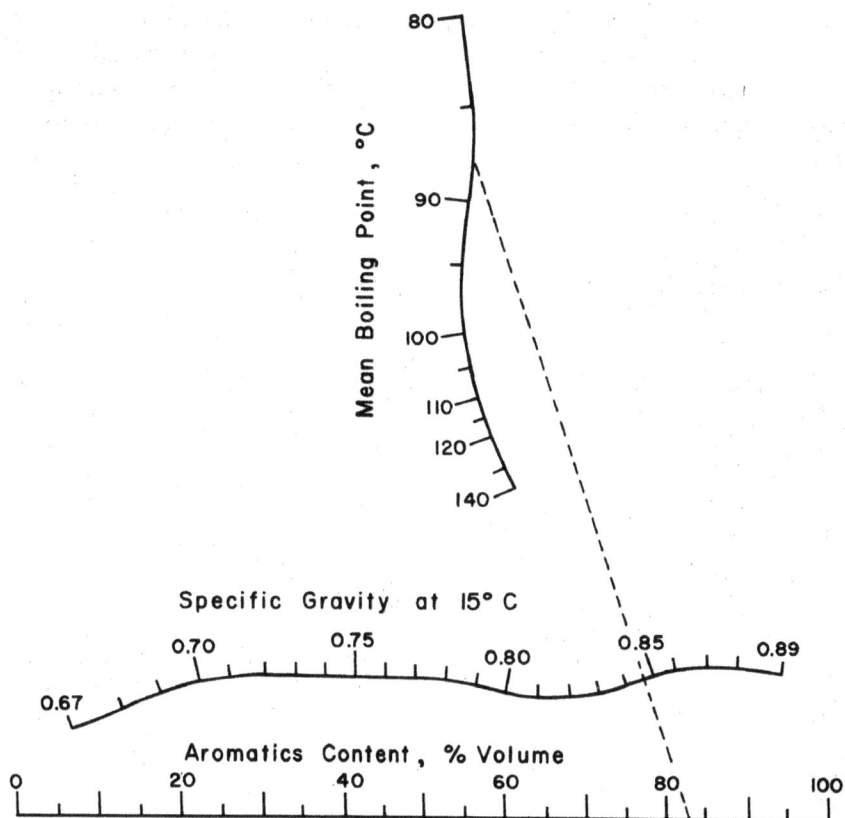

Figure 8-2

8-3   Selecting High-Temperature Gas
Bag Filters

D. B. PERLIS

Selection of equipment required to filter high-temperature air or gas is greatly simplified by using the specially prepared nomograph of Figure 8-3. The nomograph applies to bags of 11½ in. diameter. This diameter has been adopted almost universally by industry.

The nomograph permits selection of optimum collector size to suit a selected filter ratio. It automatically reconciles the variables of number of compartments, number of bags per compartment, and bag height, thus avoiding the tiresome and time-consuming trial-and-error approach.

Typical Example

To handle 100,000 ft³/min of gas at a filter ratio of 2:1, the selection is 2U5S 66-11.5-27.5.

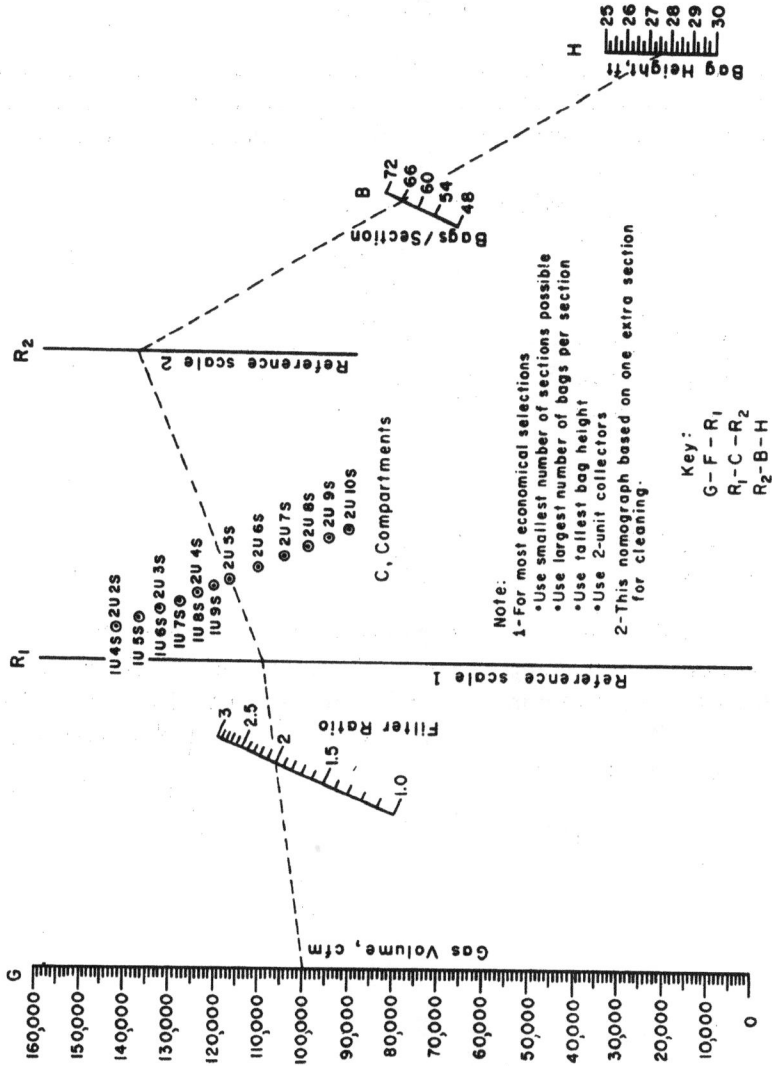

Figure 8-3

8-4   Continuous Blowdown of Boilers and Cooling Towers

F. CAPLAN

A common method of determining percentage of continuous blow-down to keep solids content constant in either boilers or cooling towers requires measurement of chloride content of makeup water and of water in boiler or cooling tower.

A chloride weight balance gives the equation

$$P = \frac{100\,Cl_m}{Cl_b}$$

where P = blowdown, % of makeup water

       Cl_m = chloride in makeup water, ppm

       Cl_b = chloride in blowdown, ppm

Typical Example

It is desired to maintain the chloride content of boiler drum water at 600 ppm. If makeup water contains 30 ppm of chloride, how much blowdown is required? On Figure 8-4 align Cl_m = 30 = 3(10) with Cl_b = 600 = 60(10), and read P = 5. This means that 5% of make-up water must be blown down.

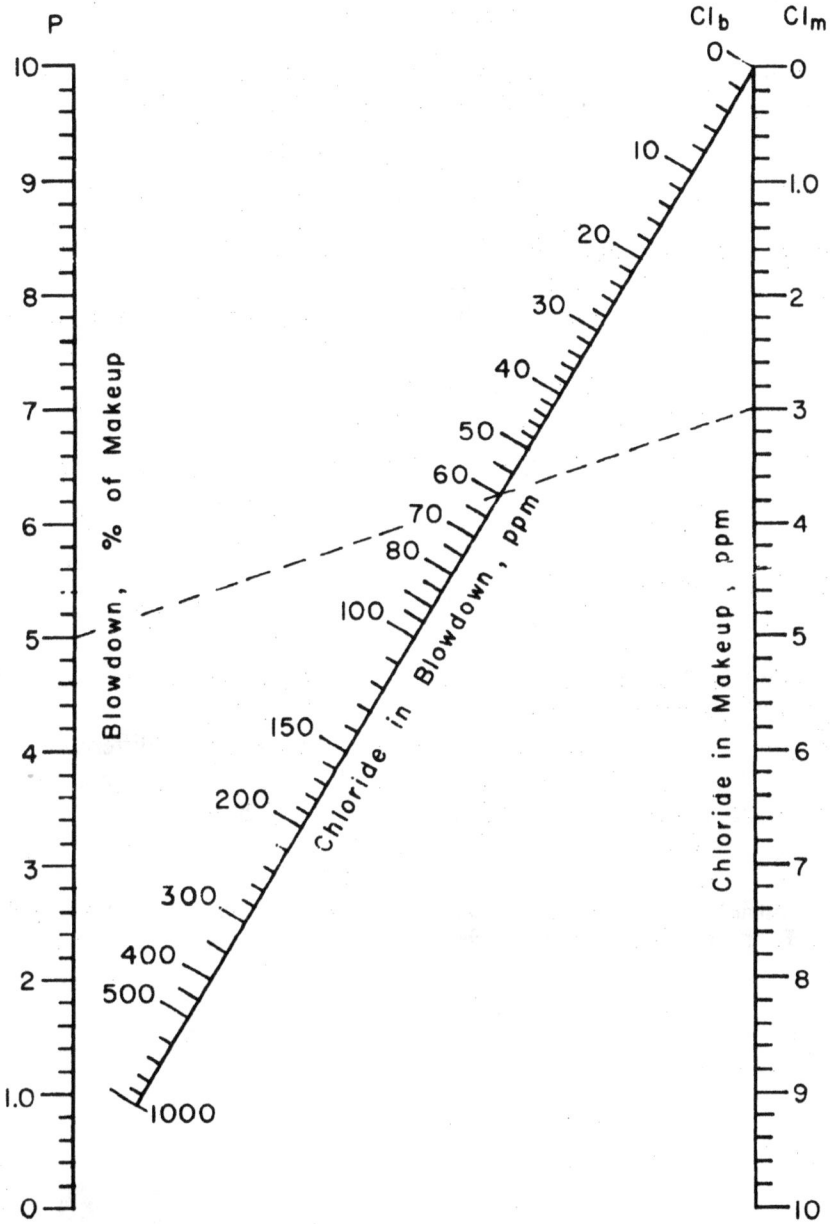

Figure 8-4

8-5   Boiling Point Correction

GERALD A. LESSELLS

Boiling points of a pure liquid near atmospheric pressure can be corrected to atmospheric pressure by the following relationship:[1]

$$\Delta P = \frac{\Delta t}{(t+273)\,(0.00010)}$$

where ΔP = difference in barometric pressure from 760 mm Hg absolute, mm Hg.

     Δt = difference in observed boiling point from atmospheric boiling point, °C

     t   = observed boiling point, °C

and ΔP and Δt have same sign.

This relationship is solved in Figure 8-5.

Typical Example

What is the normal boiling point of a liquid observed to boil at 130.0°C at 750 mm Hg absolute?

Connect 130 on the t-scale with 10 on the ΔP-scale (difference between 760 and 750). Read 0.4°C on the Δt-scale. Subtracting 0.4 from 130.0 gives a corrected boiling point of 129.6°C.

[1]Millard, E. B., *Physical Chemistry for Colleges*, 6th ed, p. 114, McGraw-Hill Book Co., Inc., New York, (1946).

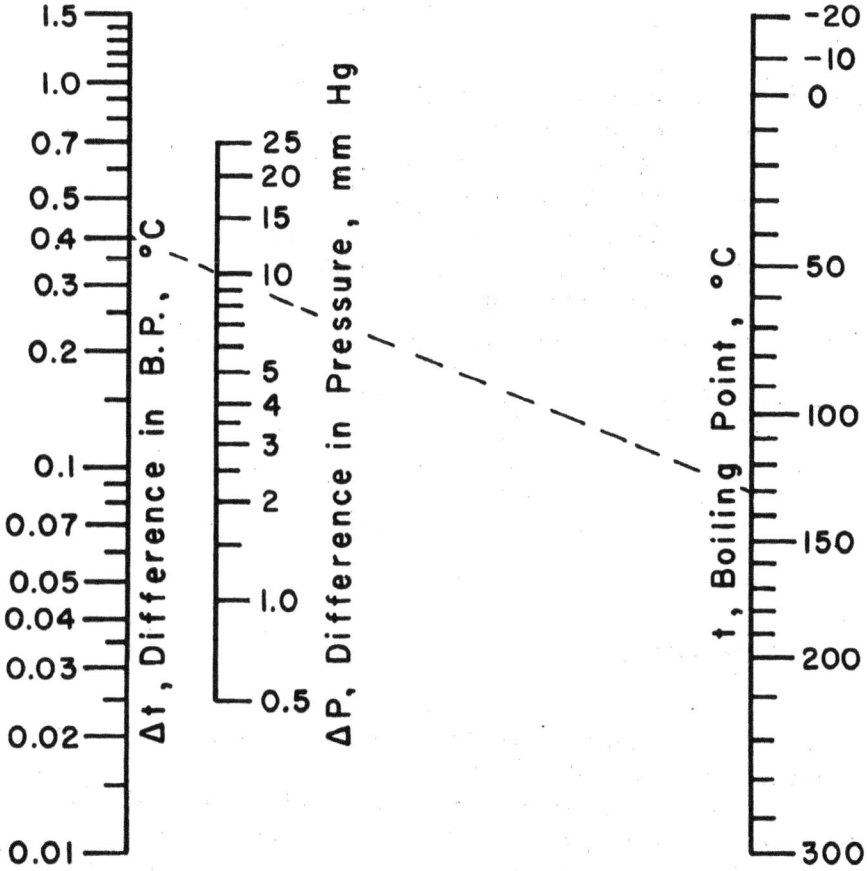

Figure 8-5

8-6   Correction of Boiling Points

B. W. HIGDON and D. S. DAVIS

For pressures near atmospheric, correction of boiling points of pure liquids to 760 mm of mercury can be made by means of the equation based on an expression given by Hoyt[1]:

$$\Delta t = \frac{(t \times 273)\, \Delta p}{K}$$

where Δt =   correction to boiling point, °C

       t   =   observed boiling point, °C (at observed pressure)

    Δp =   760 − observed pressure, mm Hg

    K   =   8000 for hydrocarbons, halogen derivatives, ethers, aldehydes, and all normal liquids; 8250 for ketones and esters; 8500 for amines; and 10,000 for water and lower alcohols

Note: Signs of Δt and Δp are always alike.

Gage Points	Compounds
1	Hydrocarbons, halogen derivatives, ethers, aldehydes
2	Ketones, esters
3	Amines
4	Water, lower alcohols

Typical Example

Use of Figure 8-6 to solve this equation is illustrated as follows: What correction should be added to the boiling point of water, observed to be 99°C at pressure of 733.3 mm of mercury, to arrive at the boiling point at pressure of 760 mm? Connect 99° on the t-scale and 4, the gage point for water, with a straight line. Connect intersection with α-axis and 760 − 733.3 or 26.7 mm on the Δp-scale with a straight line. Intersection with the Δt-scale is 1.0°C, desired correction to add to 99°C to attain 100°C, boiling point of water at one atmosphere pressure.

[1]Hoyt, C. S., *J. Chem. Ed.*, **11**, 405 (1934).

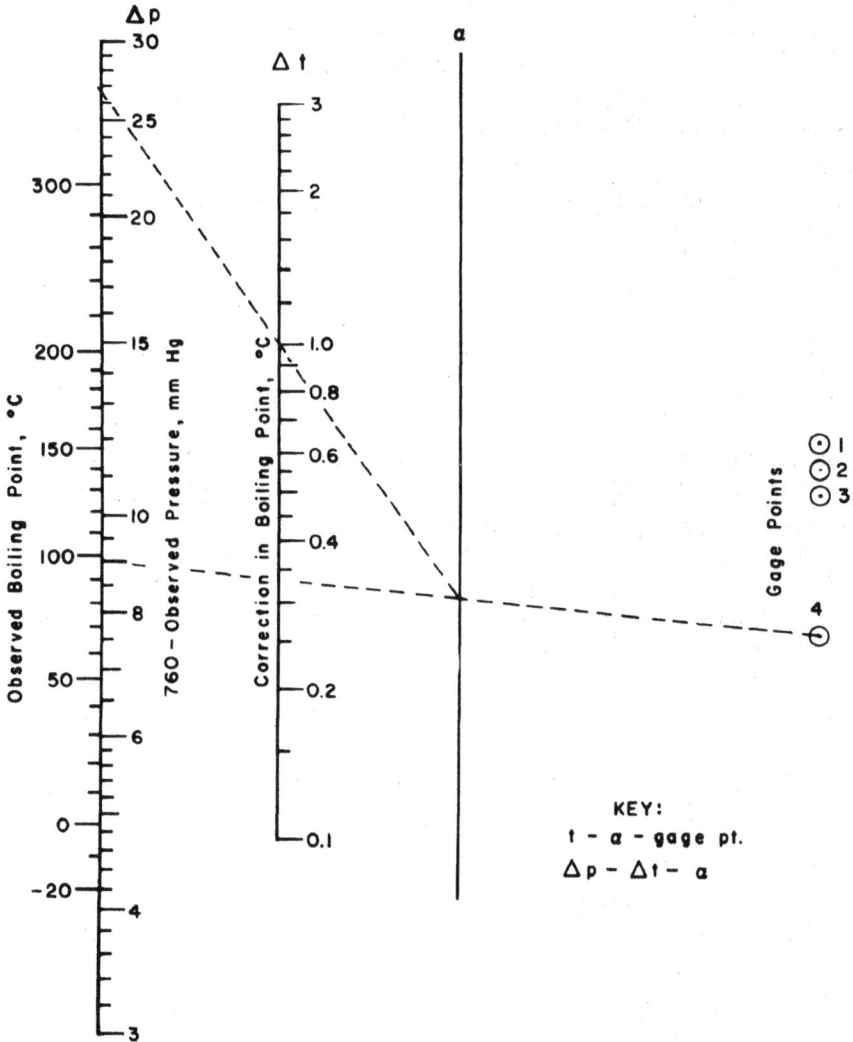

Figure 8-6

8-7   Characterization Factor of Asphalts

GEORGE E. MAPSTONE

The aromaticity of petroleum fractions is reflected in their UOP characterization factors. This property can be used to indicate the extent of treatment to which the oil has been subjected.

In the past it was not possible to determine the characterization factor for very heavy residues and asphalts. However, a recent method[1] now allows calculation of this property from specific-gravity and penetration (100 gm, 5 sec at 77°F) data. The relationship is presented in convenient nomographic form in Figure 8-7.

Typical Example

In the example, the dotted line indicates that an asphalt having API gravity of 2.0 at 60°F (specific gravity of 1.06) and a penetration of 60 has a characterization factor of 11.0.

[1]Mapstone, G. E., *Hydrocarbon Processing and Petroleum Refiner*, 40(6), 149-150, (1963).

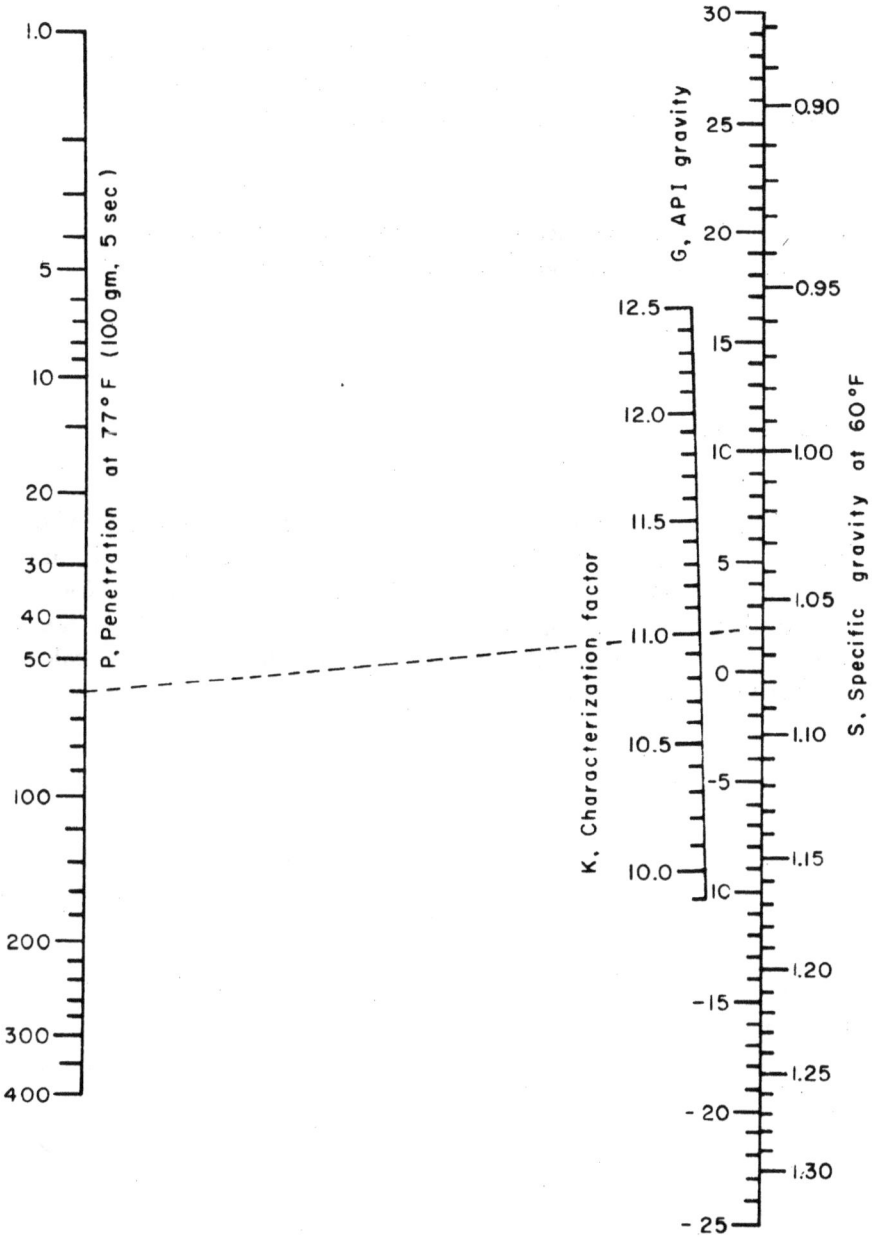

Figure 8-7

8-8   Higher Heating Value of Coal

S. E. HENRY

Figure 8-8 enables one to estimate higher heating values of coal from heating values and percentages of principal components, by means of a variant of the Dulong formula:

$$Q = 145\,C + 620\,H + 40.5\,S$$

where Q = higher heating value in Btu/lb of oven-dry coal
       C,H,S = percentages of carbon, net hydrogen and sulfur, respectively.

Typical Example

What is the higher heating value of coal when percentages of carbon, net hydrogen and sulfur are 76, 3 and 4, respectively? Connect 76 on the C-scale and 3 on the H-scale with a straight line. Note the intersection with the R-axis. Connect this point with 4 on the S-scale and read the desired higher heating value on the Q-scale as 13,000 Btu/lb of oven-dry coal.

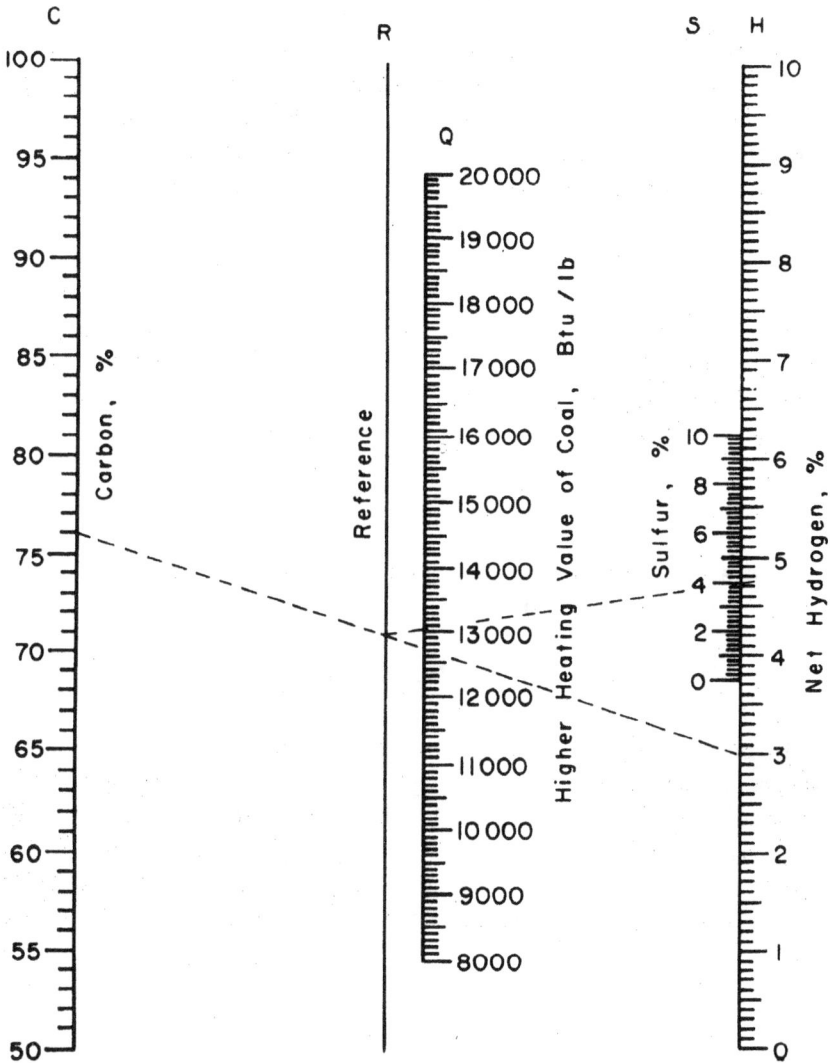

Figure 8-8

8-9   Compressor Calculations

MURRAY WOLF and D. S. DAVIS

When compressors are to be specified for any service, it is important to calculate both the hydraulic horse power (so that size of driver can be determined) and the outlet temperature of each stage (so that intercoolers and after-coolers can be specified). Temperature rise in adiabatic compression of a gas is given by the equation

$$\frac{T_2}{T_1} = \left[\frac{p_2}{p_1}\right]^{\frac{k-1}{k}} \tag{1}$$

Figure 8-9a is a nomograph based on this equation. Final temperature in °F can be established from the initial temperature (also in °F) by means of the nomograph shown in Figure 8-9b, which is based on the equation

$$t_2 = \left[\frac{T_2}{T_1}\right](t_1 + 460) - 460$$

Hydraulic horse power required to perform specific compression is given by the equation

$$HP = \left[\frac{144}{33000}\right]\left[\frac{k}{k-1}\right](p_1 v_1)\left[\left[\frac{p_2}{p_1}\right]^{\frac{k-1}{k}} - 1\right] \tag{2}$$

In performing process calculations, either the number of moles of gas being compressed or the weight of gas being compressed and its molecular weight are known. If the gas is ideal, then $p_1 v_1 = n\, RT_1$, or $p_1 v_1 = (W/M)\, RT_1$. Note that (p_2/p_1) raised to the $(k-1)/k$ power is equal to T_2/T_1; substitute for $p_1 v_1$ its equivalent value $(W/M)\, RT_1$. The equation for hydraulic horse power can then be rewritten as

$$HP = 0.000780 \left[\frac{k}{k-1}\right]\left[\frac{W}{M}\right](t_1 + 460)\left[\frac{T_2}{T_1} - 1\right] \tag{3}$$

The constant—0.000780—in equation (3) results from multiplying the constants in equation (2) by a value of R of 10.72, and dividing the result by 60. Factor 60 compensates for a change in time units, quantity W in equation (3) being expressed in lb/hr while quantity in equation (2) is expressed in ft³/min. Figure 8-9c is a nomograph based on equation (3). In conjunction with Figure 8-9a it can be used to establish the hydraulic horse power of the compressor.

Notation: HP = horse power of compressor

K = ratio of specific heat at constant pressure to specific heat at constant volume, C_p/C_v

p = pressure, $lb/in.^2$ abs.

t = temperature, °F

T = temperature, °R

W = weight of gas, lb

M = molecular weight of gas

n = moles of gas

v = volume of gas, ft^3/min

Subscript 1 = initial conditions

Subscript 2 = final conditions

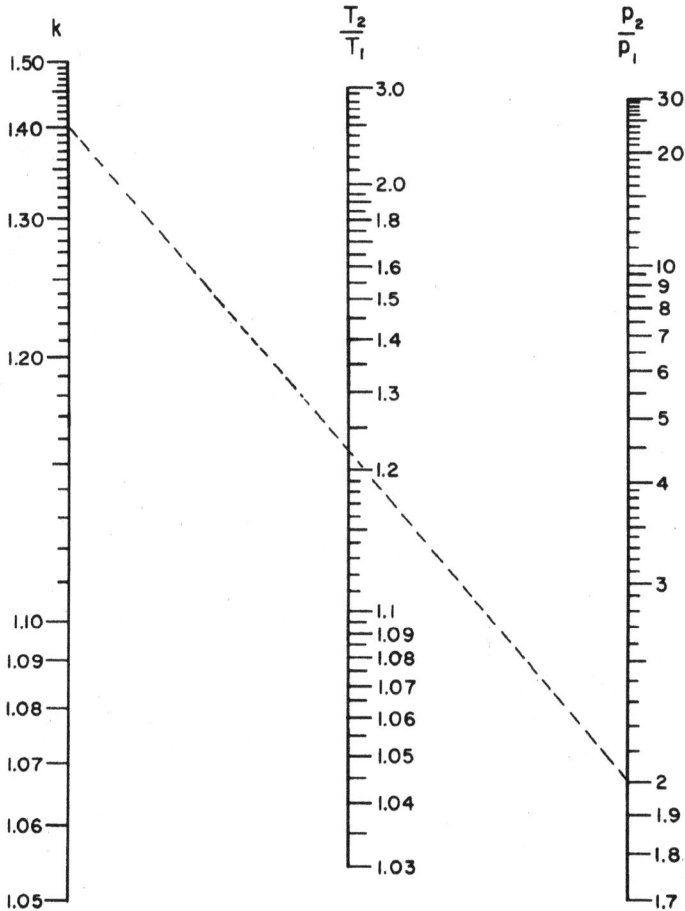

Figure 8-9a

Typical Example

What would be the hydraulic horse power required to compress 2000 lb/hr of air (molecular weight of 29 and k of 1.4) at initial temperature of 90°F from 14.7 to 29.4 lb/in.² abs.? What is the temperature of the air leaving the compressor?

Following the broken index line on Figure 8-9a, join 1.4 on k-scale with 2 on the p_2/p_1-scale. Read the ratio of final to initial absolute temperature (1.22) on the T_2/T_1-scale. To determine the final temperature in °F, follow the broken index line on Figure 8-9b and join 90 on the t_1-scale with 1.22 on the T_2/T_1-scale. Read 210°F on the

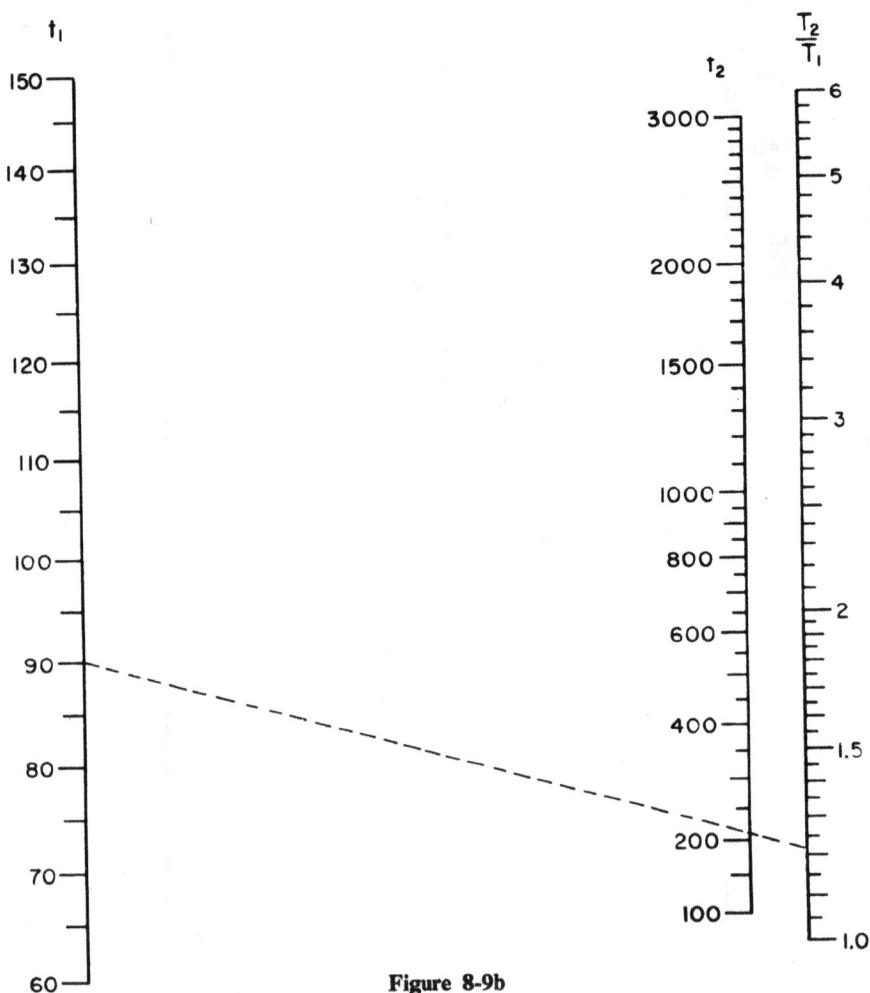

Figure 8-9b

t_2-scale.

To determine the horse power, join 1.4 on the k-scale of Figure 8-9a with 90 on the t_1-scale and note the intersection on the α-axis. Join this point with 1.22 on the T_2/T_1-scale and note the intersection with the β-axis. Now join 2000 on the W-scale with 29 on the M-scale and note the intersection with the n-scale, 69. Join 69 on the n-scale with the point on the β-axis previously noted; read horse power, 22.6, on HP scale.

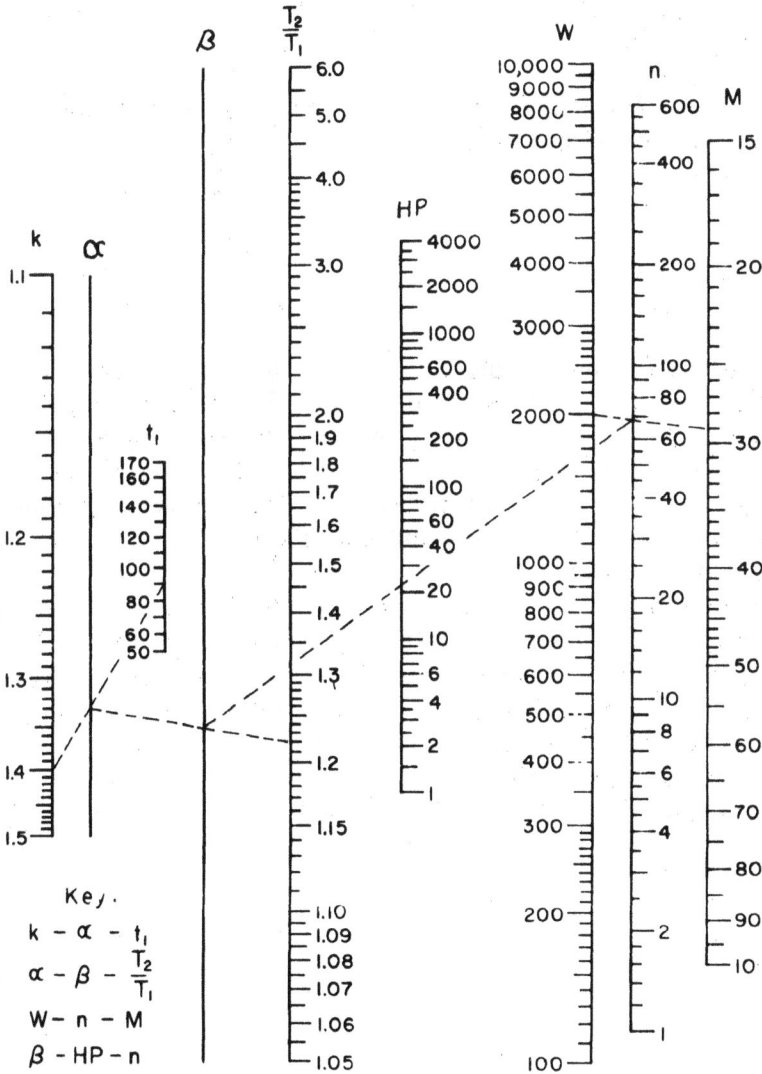

Figure 8-9c

8-10   Concentration and Dilution

J. G. LOWENSTEIN and D. S. DAVIS

The weight of solute W_e that must be added to P weight units of solution in order to increase the concentration from C_P to $C_H\%$ equals $P(C_H - C_P)/(100 - C_H)$. The weight of solvent W that must be added to P weight units of solution to decrease the concentration from C_P to C_L per cent is equal to $P(C_P - C_L)/C_L$. Any consistent weight units may be employed.

These equations can be solved readily and accurately by means of Figure 8-10. For weights outside the range of the chart, multiply all weight units by a convenient constant.

Typical Examples

1) How much solute should be added to 80 lb of solution to increase the concentration from 35 to 60%? Following the key for concentration (above the slant axis) and the dotted illustrative lines, connect $C_H = 60$ and $C_H - C_P = 60 - 35 = 25\%$ with a straight line, and mark the intersection with the slant axis. Connect this point and P = 80 with a straight line and note the intersection with the W_e-scale at 50 lb of solute.

2) How much solvent should be added to 800 g of solution to decrease the concentration from 65 to 40%? Following the key for dilution (below the slant axis and, in this case, the same dotted illustrative lines), connect $C_L = 40$ and $C_P - C_L = 65 - 40 = 25\%$ with a straight line, and mark the intersection with the slant axis. Connect this point and P = 80 (for 800 g) with a straight line and read the desired value on the W_t scale as 50 (for 500 g) of solvent.

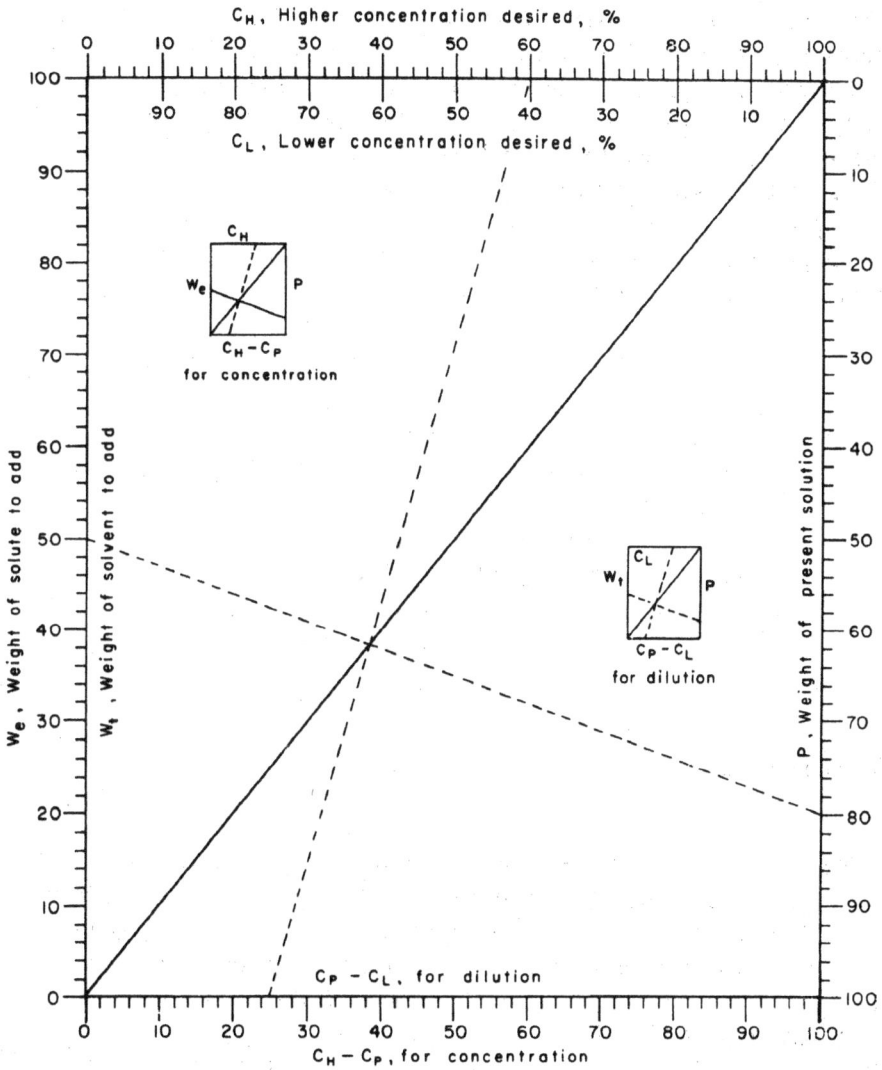

Figure 8-10

8-11 Water Requirement for Barometric Condensers

JAVIER F. KUONG

In the sizing of barometric condensers to handle a given condensable vapor load, it is often necessary to calculate the amount of water necessary to accomplish condensation of vapors. If the vapor load is known, the heat load can be estimated from a knowledge of the "approach to saturation." This is defined as the difference between the saturation temperature of the vapor and the temperature of the water at the discharge from the condenser.

A knowledge of the total heat load and the temperature rise of the water permits calculation of the flow rate of water required. Figure 8-11 gives the answer directly and is based on the following equation[1]:

$$G = \frac{[\lambda + (t_s - t_o)]\, W}{500\, (t_o - t_w)}$$

where G = water flow rate, gal/min

λ = latent heat of evaporation of saturated vapor, Btu/lb of vapor

t_o = water discharge temperature, °F

t_s = saturation temperature of vapor, °F

t_w = inlet water temperature, °F

W = weight rate of vapor to be condensed, lb/hr

Typical Example

Vapor load is 10,000 lb of saturated steam per hour at 27 in. of vacuum. Inlet water temperature is 70°F. Desired "approach to saturation" is 5°F ($t_s - t_o$). What is the required water flow rate?

The temperature of saturation of steam at a vacuum of 27 in. of Hg is $t_s = 115°F$ and its latent heat is 1027 Btu/lb (obtained from any standard handbook). As the approach to saturation is 5°F = ($t_s - t_o$) = 115°F − t_o, $t_o = 110°F$. Therefore the temperature rise is $\Delta t_R = 110 - 70 = 40°F$. Also, the total heat load per lb of vapor is 1027 + 5 = 1032 Btu/lb.

Connect 1032 on the extreme left-scale with 10,000 on the W-scale to intersection with the R-line. With this intersection as a pivot point, connect with 40 on the Δt_R-scale and read 516 gal/min as the water flow rate.

[1]Perry, J. H., *Chemical Engineers' Handbook*, 3rd ed., McGraw-Hill Book Co., New York, 1950.

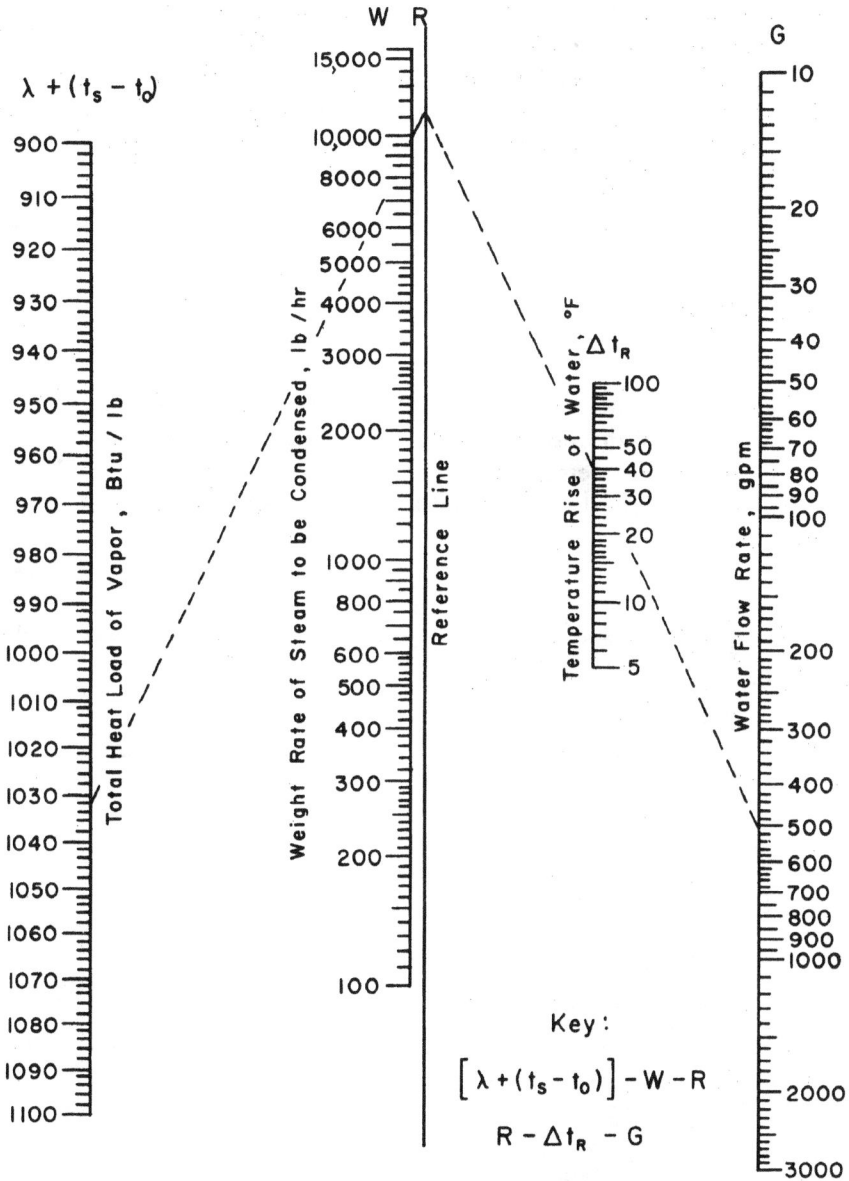

Figure 8-11

8-12   Thermal Conductance of Air Spaces

GEORGE E. MAPSTONE

There appears to be no simple method for calculating thermal conductances of narrow air spaces which often are used for insulation purposes in building construction. The experimental data of Rowley and Algren[1] is presented here in a convenient nomographic form, Figure 8-12. The chart has of the same order of accuracy as the original tabulated data and allows simpler, quicker and more accurate interpolation.

Typical Example

What is the thermal conductance of an air gap 0.500 in. wide between plates if the mean temperature of the plates is 110°F? Connect 110 on the temperature scale with 0.500 on the width scale. This line cuts the thermal-conductance (center) scale at 1.52 Btu(ft)/ (hr) (ft²) (°F).

---

[1]Rowley, F. B., and Algren, A. B., *Trans. ASHVE*, 39, (1935); data tabulated by O. W. Esbach, *Handbook of Engineering Fundamentals*, John Wiley and Sons, New York, 12-33, 1936.

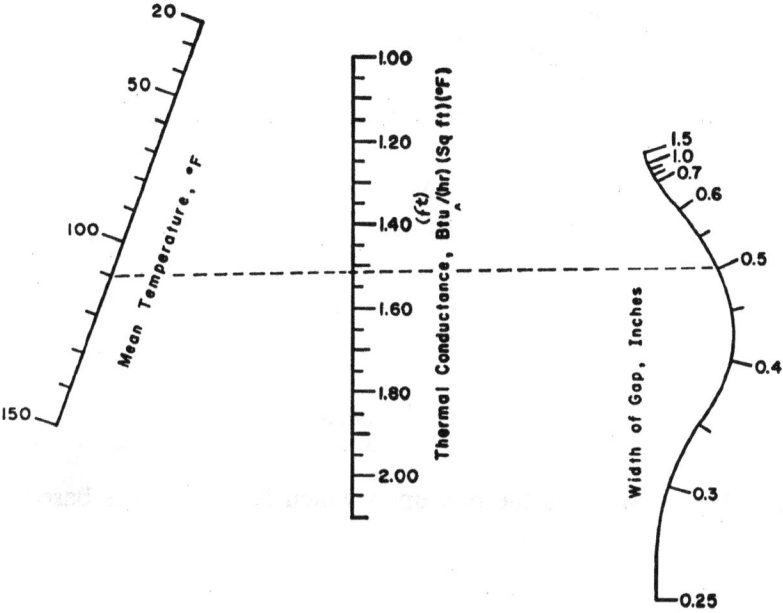

Figure 8-12

8-13 Volume of a Cone

C. W. MATTHEWS

Conical shapes are encountered in pump sumps, agitators, tanks, and many stock-piles.

The volume for a right circular cone is

$$V = \frac{\pi R^2 a}{3} \tag{1}$$

where V = volume
   R = radius of the base
   a = altitude

When ϕ is the angle between an element and the base, then

$$\tan \phi = \frac{a}{R} \tag{2}$$

so that

$$a = R \tan \phi \tag{3}$$

and

$$V = \frac{\pi R^3 \tan \phi}{3} \tag{4}$$

The last equation is the one upon which Figure 8-13 is based.

Typical Example

The broken index line shows that the volume of a right circular cone with a base diameter of 48 ft and an angle ϕ of 40° is 12,100 ft³.

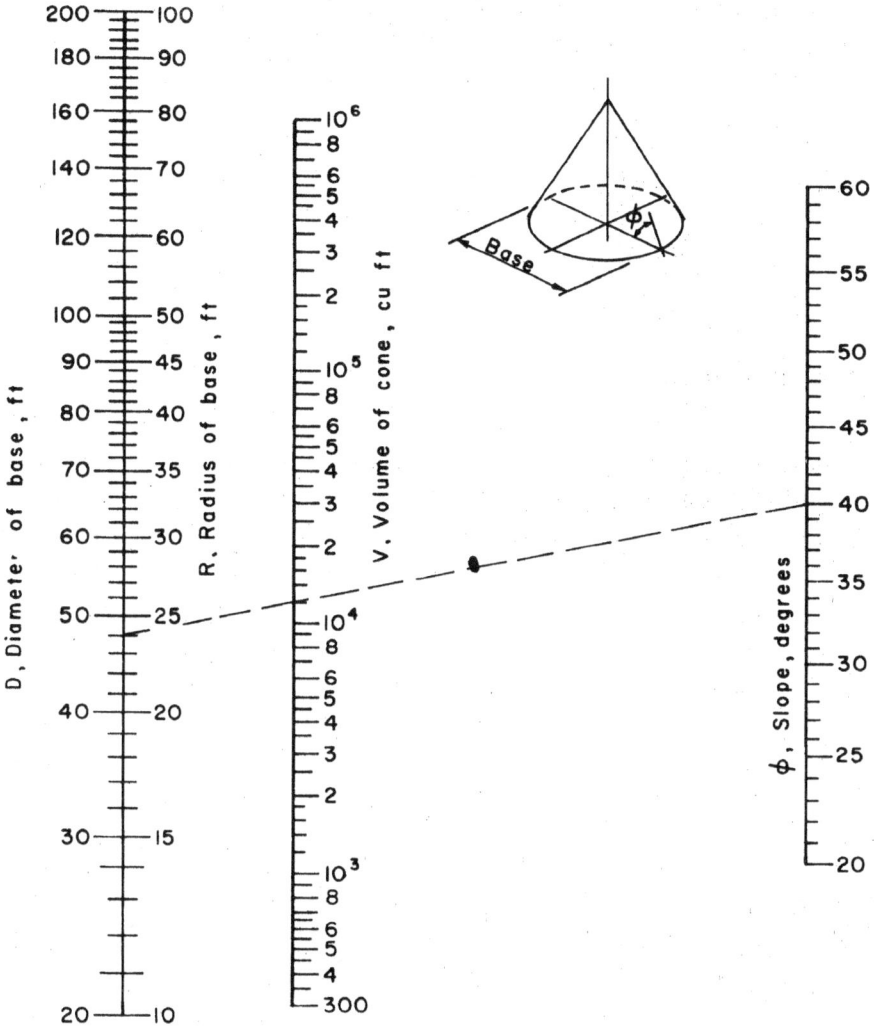

Figure 8-13

8-14   Conveyor-Belt Production

J. B. SHINAL

Figure 8-14 aids in scheduling speeds of conveyor belts for optimum efficiency. It is based on the equation

$$v = \frac{Pd}{60}$$

where   v   = speed of conveyor belt, in./min
        P   = hourly production rate, pallets
        d   = distance between pallets, in.

Typical Example

On the chart, the broken index line shows that the correct speed for a conveyor belt is 20 in./min when the distance between pallets is 24 in. and the hourly production rate is 50 pallets.

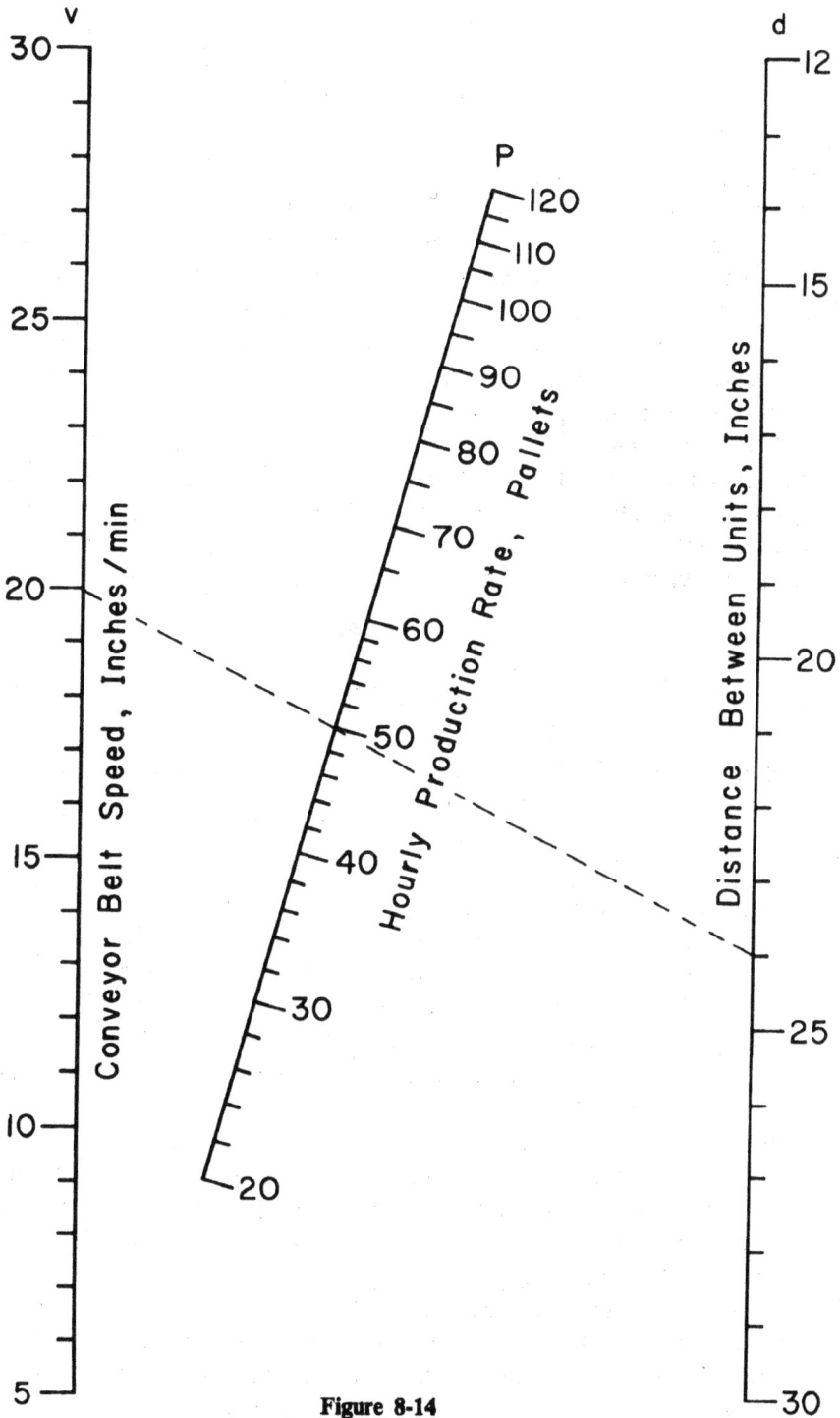

Figure 8-14

8-15 Fractional Crystallization

JOSEPH I. LACEY and D. S. DAVIS

Some mixtures can be purified by partially crystallizing one of the components under conditions of controlled temperature. Phases are then separated in a centrifuge. Some of the impurities remain with the crystals as mother liquor, wetting the surfaces. Some of the solid phase remains with the liquid phase (mother liquor). This discussion is limited to non-aqueous systems, such as might be described as crystallizing from a "melt." Typical products include phenol, naphthalene and benzene.

From material balances, equations[1] have been developed that relate the purity and quantity of feed stock, mother liquor, and impure crystals. Assuming that freezing-point lowering is proportional to concentration of impurities, the authors' final equations become, in terms of weight percentages,

$$\Delta P = \frac{E \Delta F}{10 - 10W(1-E)} \tag{1}$$

$$E = \frac{100 \, \Delta P}{\Delta L} \tag{2}$$

Substitution of equation (2) into equation (1) yields

$$W = \frac{100 \, (\Delta L - \Delta F)}{\Delta L - \Delta P} \tag{3}$$

where ΔP = freezing-point lowering of product, °C
 ΔF = freezing-point lowering of feed, °C
 ΔL = freezing-point lowering of mother liquor, °C
 E = weight percentage of mother liquor on crystals, referred to total weight of impure crystals
 W = weight percentage of impure crystals, referred to total feed

E represents the total mother liquor on the solid crystals; hence, $100\% - E$ is a measure of the efficiency of the separation. The purity of the crystals is proportional to ΔP and is a measure of the total desired product (solid phase plus the quantity dissolved in the mother liquor on the crystals).

One must observe the freezing-point lowerings of the mother liquor, crystal product, and feed. Freezing points of these materials are determined; then each is subtracted from the freezing point of the pure compound.

Typical Example

The use of Figure 8-15 is illustrated as follows: A certain benzol stream has a crystal point of 0.93°C. It is fractionally crystallized and centrifuged to yield a crystal with a crystal point of 4.98°C and a mother liquor with a crystal point of −0.32°C. The crystal point of the pure benzene is 5.53°C. The freezing-point lowering for the feed crystals is 4.60°C; for mother liquor, 5.53°C to −0.32°C or 5.85°C; and for the product, 5.53°C −4.98°C or 0.55°C. What percentage of mother liquor is retained on the crystals, and what is the percentage yield of impure crystals?

Following the key, connect 5.85 on the ΔL_1-scale and 4.60 on the ΔF-scale with a straight line; mark the intersection with the α-axis. Then connect 5.85 on the ΔL_2-scale and 0.55 on the ΔP-scale with a straight line, marking the intersection with the β-scale and noting the intersection with the E-scale as 9.4%, the percentage of mother liquor retained on the crystals. Connect the intersections on the α and β-scales with a straight line, and read the weight percentage of impure crystals as 23.6% on the W-scale.

[1]Molony and Roberts, *J. Applied Chem.*, **11**, 283, 1961.

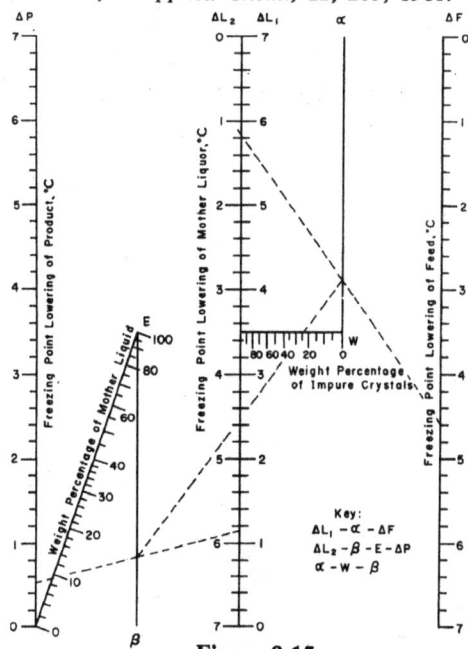

Figure 8-15

8-16   **Thicknesses of Electrodeposits**

J. E. CHILTON, E. F. DUFFEK, and D. S. DAVIS

Figure 8-16 permits rapid estimation of thicknesses of electrodeposits. It is based on the equation

$$L = \frac{2.63 \cdot 10^{-4} \cdot I\,\theta \text{ (atomic weight)}}{D \text{ (valence change)}}$$

where L = thickness of deposit, .001 in.
       I   = current density, amp/ft²
       θ   = plating time, min
       D = density of deposit, g/cm³

Typical Example

What thickness of cadmium can be deposited electrolytically in 15 min at a current density of 24 amp/ft² if the current efficiency is 100%? Following key, connect 24 on the I-scale and 15 on θ-scale with a straight line. Note the intersection with α-axis. Connect this point and the point for cadmium on the metal scale with a straight line. Read the thickness of deposit as 0.61 of .001 in. on the L-scale. Multiplication of this theoretical thickness by current efficiency as a fraction gives actual thickness.

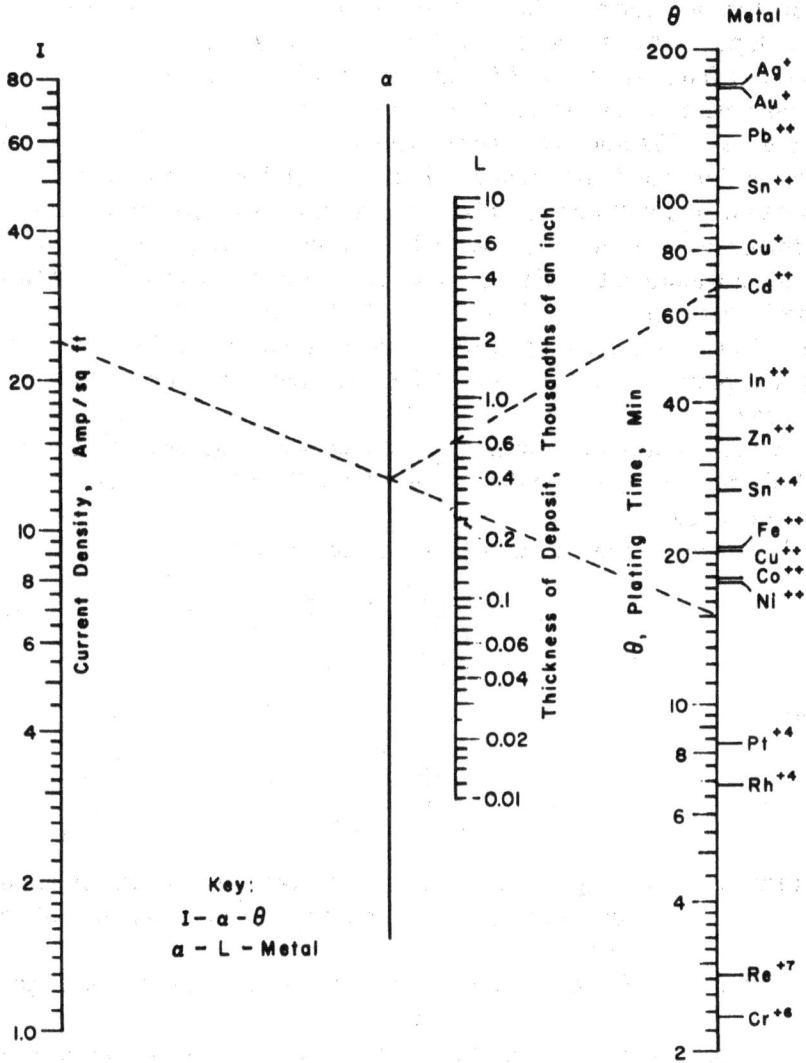

Figure 8-16

8-17   Contaminants from Flare Stacks

P. D. SHROFF

In the operation of flare stacks, toxic or odor hazards existing at breathing levels can be due to (a) unburned waste gases present because of flame failure and (b) gas-combustion products present when atmospheric-dispersion conditions are poor.

Figures 8-17a and 8-17b permit quick determination of the distance from a flare stack at which maximum ground-level concentrations of contaminants will occur. They are based on so-called moderate air-turbulence conditions, associated with an unstable atmosphere in which air temperature decreases relatively rapidly with height (atmospheric-dispersion coefficient, P, is 0.10). If so-called average air turbulence — associated with a neutral atmosphere (P = 0.05) — should prevail, the distance obtained via the nomographs should be doubled.

Figure 8-17a is to be used when U_c, critical wind velocity, is under 1.5 ft/sec. If U_c is equal to or greater than 1.5, Figure 8-17b must be employed.

The nomographs are based on the equations[1]

$$L = \frac{1.052 \times 10^3 \, D/P}{c/f} \tag{1}$$
$$\text{(for } U_c \geq 1.5)$$

and

$$L = \frac{150}{P}\left[\frac{Q_t}{c/f}\right]^{0.5} \tag{2}$$
$$\text{(for } U_c > 1.5)$$

where L = distance from the flare stack at which maximum ground-level concentrations of contaminants will occur, ft

       f = Q_m/Q_t = fraction of potentially toxic waste gases present

      C = average ground-level concentration of waste gases that occurs over a period of 30 min or longer downwind of a source, ppm by volume (Peak 1-min concentrations will be about 10 times C)

     Q_t = total gas-emission rate at temperature t, ft³/sec

    Q_m = Contaminant-emission rate at atmospheric temperature and pressure, ft³/sec

t = Contaminant gas temperature, °K. For gases at
atmospheric pressure, t = T(molecular weight of
stack gases)/29.

T = Absolute ambient air temperature, °K

D = Flare-tip ID, in.

The critial wind velocity, U_c, may be calculated from the equations

$$U_c = 16.7 \times 10^{-5} \, V \left[\frac{c}{f} \right]$$ (3)

and

$$V = 183.2 \, \frac{Q_1}{D^2}$$ (4)

where V=gas exit velocity in ft/sec.

Typical Examples

1) Assume Q_t = 39.3 ft³/sec, f = 0.5, C = 15 ppm, D = 6 in.
and V = 200 ft/sec, where U_c is less than 1.5 ft/sec. How far from
the stack will contaminants occur at maximum concentration, under
conditions of both moderate and average air turbulence?

Figure 8-17a will be used. Draw a straight line from 0.5 on the
f-scale to 15 on the C-scale, marking the intersection on the reference
line R. Next draw a line from 39.3 on the Q_t-axis through this intersection point, and extend the line on the L-scale to find a distance
of 1725 ft. This is the value for moderate turbulence. For average
air turbulence, the value would be twice that, or 3450 ft.

2) Assume the same conditions as above, except that C = 25 ppm.
In this case, U_c turns out to be 1.67 ft/sec, and Figure 8-17b is applicable. Starting out as before, a line is drawn through the R-scale
connecting 0.5 on the f-scale and 25 on the C-scale. A second line is
then constructed through the intersection at R, connecting 6 on the
D-scale with 1260 ft, on the L-scale. Again, for average air turbulence, this figure is doubled to yield 2520 ft.

Note: The equations and nomographs are only applicable for a
gas-exit velocity, V, greater than 80 ft/sec.

[1]F. T. Bodurtha, Jr., "Flare Stacks—How Tall?" *Chemical Engineering*, **65**,
pp. 177-180, Dec. 15, 1958.

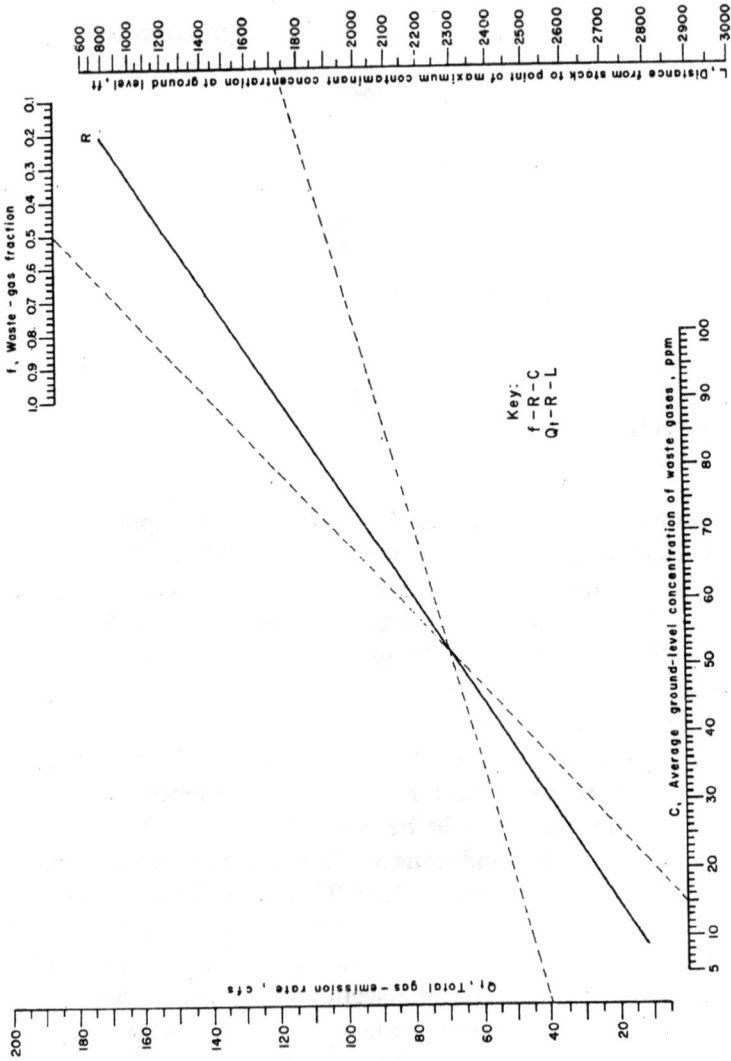

Figure 8-17a

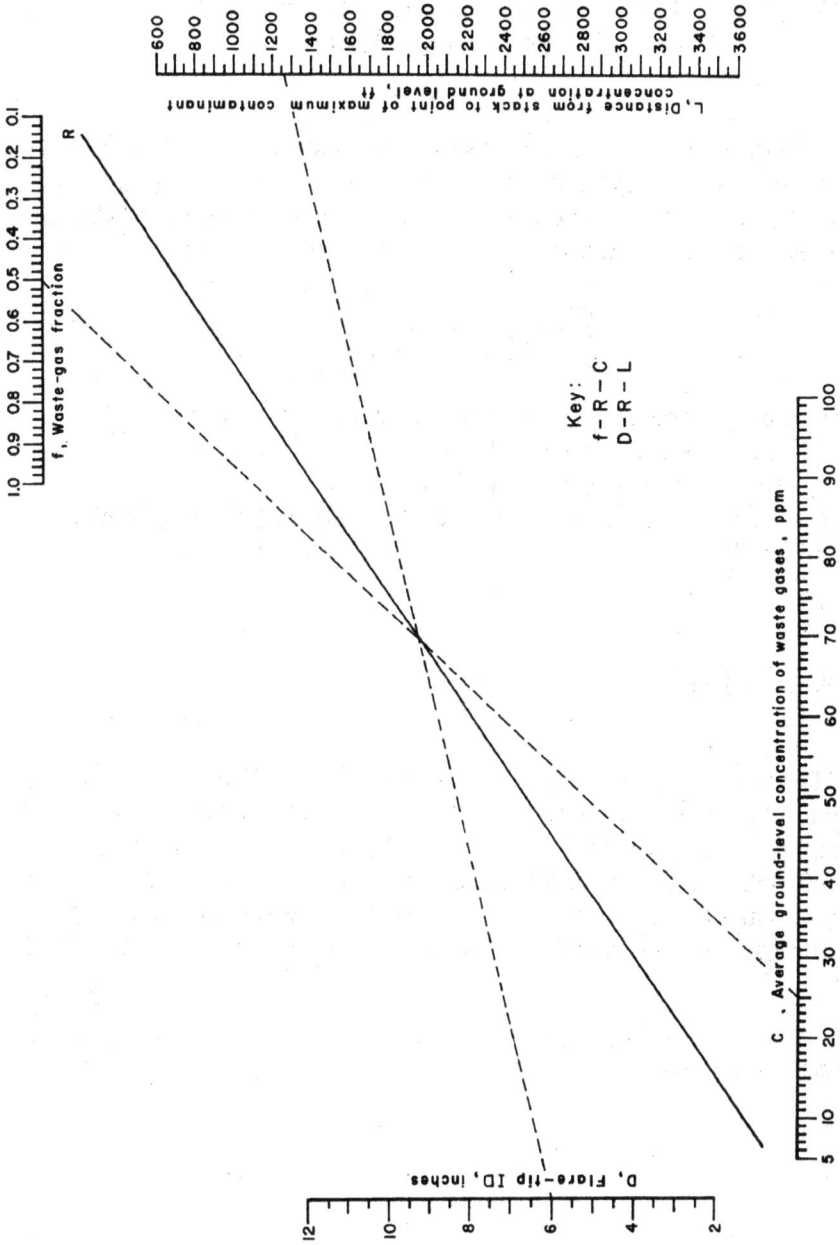

Figure 8-17b

8-18   **Friction-Correction Factors for**
                          Fracturing Fluids

D. S. DAVIS

Friction pressures for suspensions of sand in turbulent flow, as read from charts,[1] must be multiplied by correction factors that depend upon the density of the fluid, which is a function of the sand content. These variables have been correlated in the expressions

$$d = a + b \log (S + 10)$$
$$F = 0.10 \, d + 0.20$$
$$F = 0.2 + 0.10 \, [a + b \log (S + 10)]$$

where d = density of the sand suspension, lb/gal
        S = sand content, lb/gal
        F = friction-correction factor

These equations can be solved readily and accurately by means of Figure 8-18.

Typical Example

The use of the chart is illustrated as follows: What is the density of a sand suspension and what friction-correction factor should be employed when the sand content is 4.80 lb/gal of fluid and the gravity is 30° API? Connect 4.80 on the S-scale and 30 on the gravity scale with a straight line and note the intersection with the density scale at 9.90 lb/gal and the scale for friction-correction factor at 1.19.

[1]*Calculations for Friction Loss in Fracturing Fluids*, Halliburton Oil Well Cementing Co., Duncan, Okla.

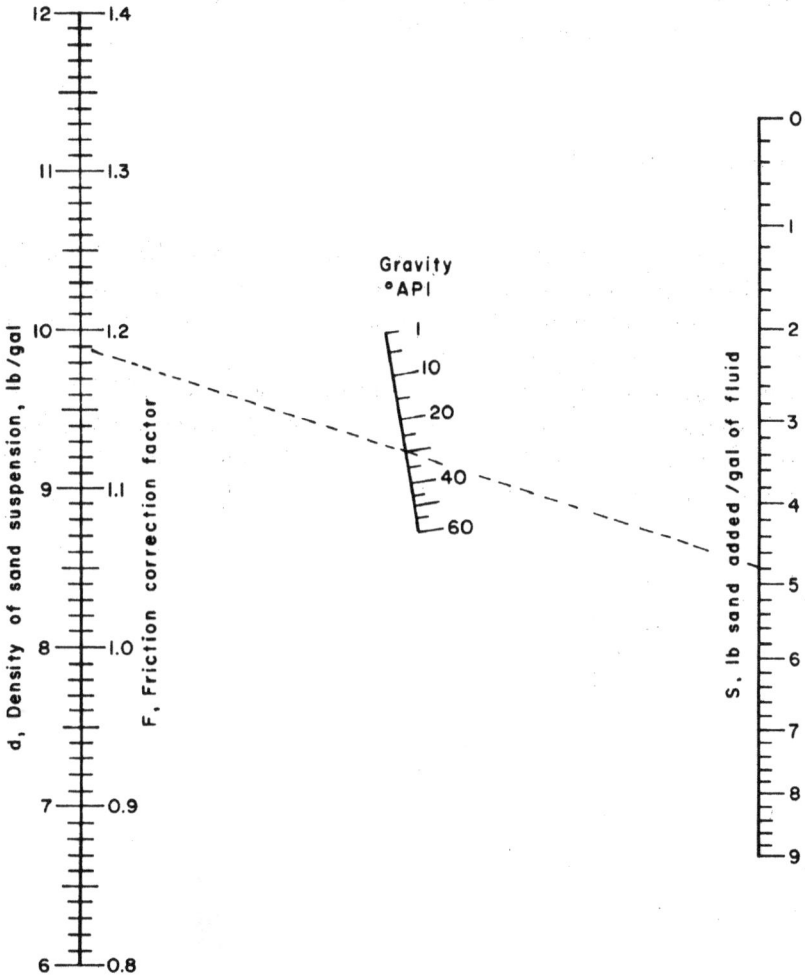

Figure 8-18

8-19   Injection of Inhibitor

H. A. RYAN

When using corrosion inhibitors in a process, it is frequently necessary to vary inhibitor injection rates and dosages because of changes in operating conditions, such as throughput. With Figure 8-19, such adjustments can be calculated quickly.

Typical Example

It is desired to inject 6 ppm of inhibitor into a 10,000-bbl/day stream by means of a 1 gal/hr injection pump (42-gal barrels are understood). Find how many gallons of inhibitor will be required per day and what the dilution should be.

Draw a straight line from 6 ppm on the A-scale to 10,000 bbl/day on the B-scale. Read required inhibitor injection rate of 2.5 gal/day on the C-scale. Connect this point with 1 gal/hr on the E-scale. Read a dilution of 10.5% on the D-scale.

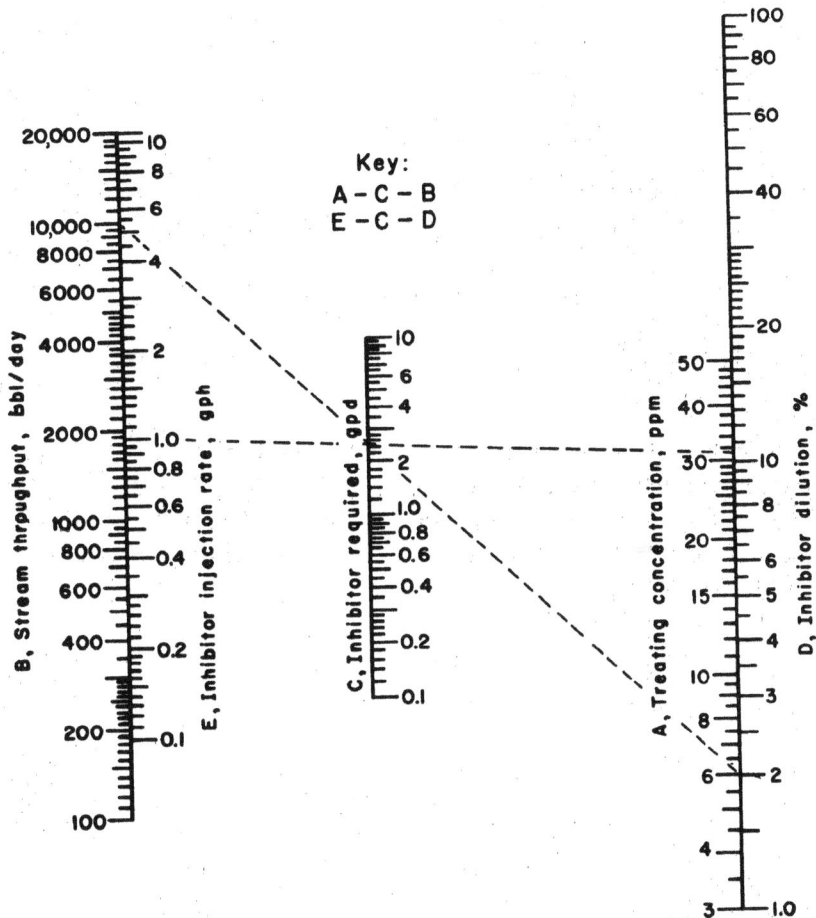

Key:
A – C – B
E – C – D

Figure 8-19

8-20   Temperature of Mixtures

F. CAPLAN

Figure 8-20 can be used to determine the temperature of a mixture of two gases or liquids of the same density. It is based on the equation

$$(W_A)(C_p)(t_A - t_{A+B}) = (W_B)(C_p)(t_{A+B} - t_B)$$

or

$$(t_{A+B} - t_B) = \left[\frac{W_A}{W_A + W_B}\right](t_A - t_B)$$

where W_A and W_B = amounts of gases or liquids to be mixed, lb
        t_A and t_B = temperatures of gases or liquids, °F

Typical Example

If 300 lb of air at 10°F is mixed with 450 lb of air at 35°F, what is the temperature of the mixture? Align W_A = 450 with $(t_A - t_B)$ = $(35 - 10)$ = 25. Next, align intersection of this line at R-scale with $(W_A + W_B)$ = 750. Read $(t_{A+B} - t_B)$ = 15. Therefore, t_{A+B} = 10°F + 15°F = 25°F.

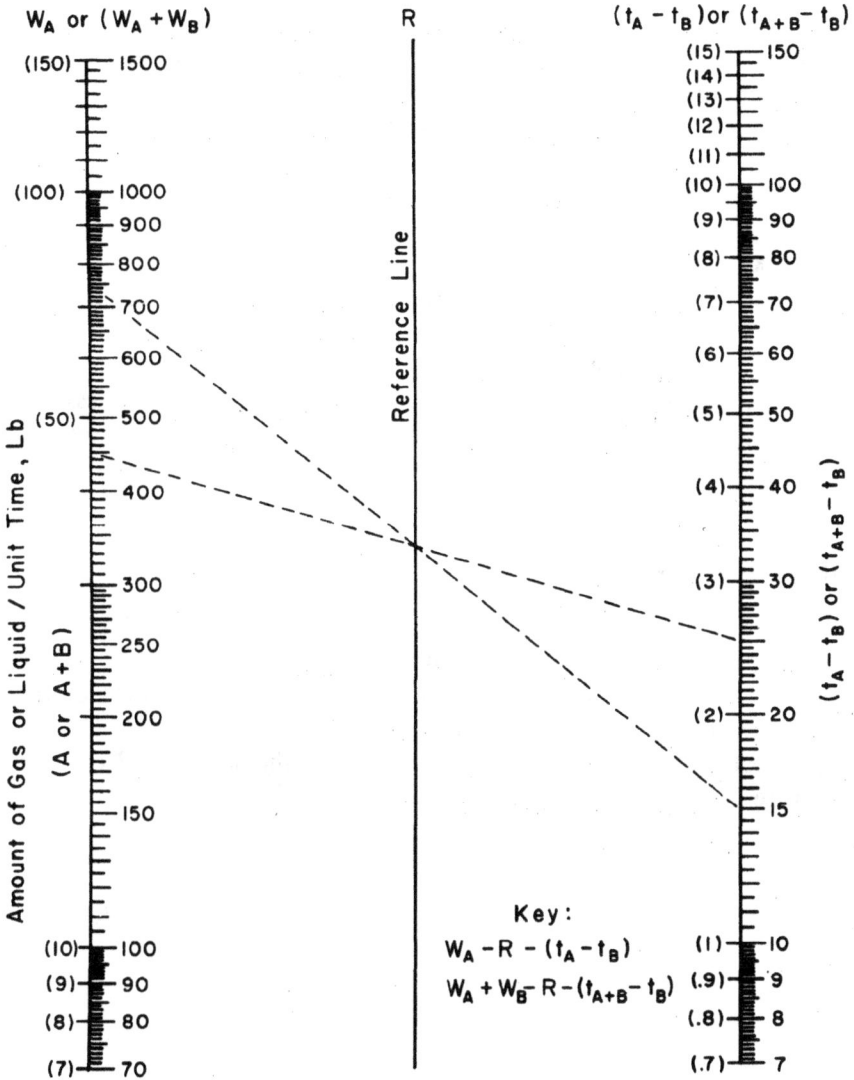

W_A or ($W_A + W_B$)

(150) — 1500

(100) — 1000
— 900
— 800
— 700
— 600
(50) — 500
— 400
— 300
— 250
— 200
— 150
(10) — 100
(9) — 90
(8) — 80
(7) — 70

Amount of Gas or Liquid / Unit Time, Lb
(A or A+B)

R

Reference Line

($t_A - t_B$) or ($t_{A+B} - t_B$)

(15) — 150
(14)
(13)
(12)
(11)
(10) — 100
(9) — 90
(8) — 80
(7) — 70
(6) — 60
(5) — 50
(4) — 40
(3) — 30
(2) — 20
— 15
(1) — 10
(.9) — 9
(.8) — 8
(.7) — 7

($t_A - t_B$) or ($t_{A+B} - t_B$)

Key:
$W_A - R - (t_A - t_B)$
$W_A + W_B - R - (t_{A+B} - t_B)$

Figure 8-20

8-21   Percentage Moisture in Steam

JAMES K. O'HARA

Figure 8-21 was constructed empirically from the Keenan-Keyes steam tables. It provides an easy and quick method for determining the percentage of moisture in steam when measured by a throttling calorimeter. The chart is based on the following equation:

$$M = \left[1 - \frac{(h - h_f)}{h_{fg}} \right] 100$$

where   M = % moisture
   h = enthalpy of the throttled steam at atmospheric pressure (determined from the resulting steam temperature at atmospheric pressure)
   h_f = enthalpy of saturated water at the initial steam pressure
   h_{fg} = latent heat of evaporation at the initial steam pressure

Typical Example

Determine the percentage moisture in steam if the steam pressure ahead of the calorimeter is 200 lb/in.^{2} abs. The temperature of the steam leaving the throttling calorimeter is 260°F.

Connect 200 on the P-scale with 260°F on the T-scale. Read the moisture content on the M-scale as 2.9%.

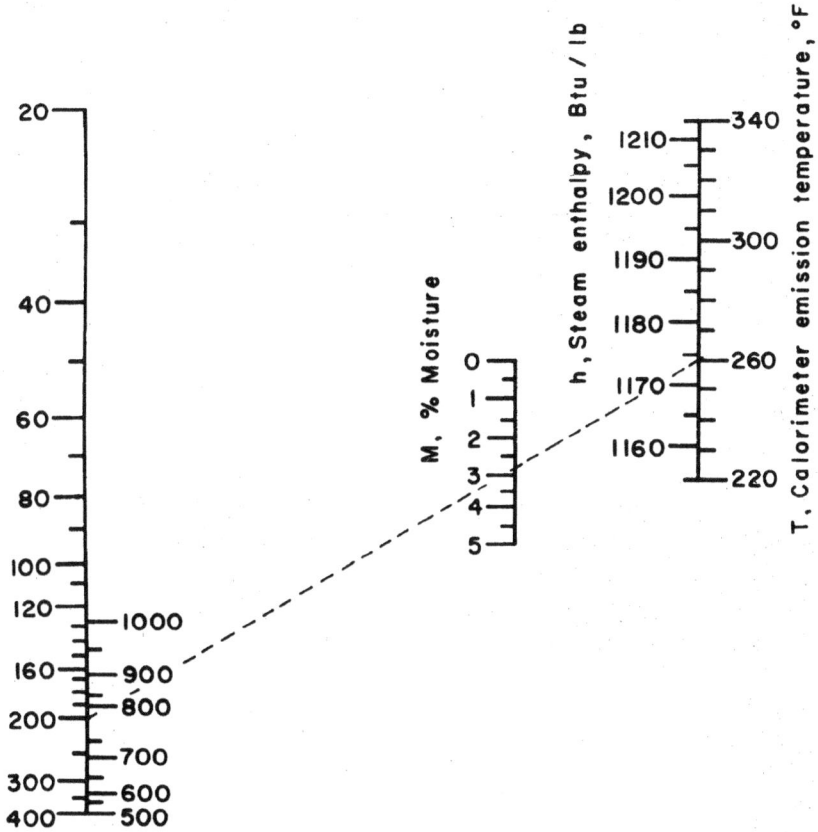

Figure 8-21

8-22   Scaling-up Equipment Costs

WILLIAM RESNICK and D. S. DAVIS

Those who estimate the cost of chemical equipment and plants are frequently faced with the problem of determining costs at a capacity level, for which there are no data.

If, however, there is cost data at some other capacity level, they can estimate cost at the desired capacity level with Figure 8-22.

The chart is based on the equation

$$S_a/S_b = (C_a/C_b)^n$$

where S_a and S_b = costs of units a and b

$\quad\quad$ C_a and C_b = capacities of units a and b

$\quad\quad\quad\quad$ n = cost scale-up factor.

Values of n for different equipment[1,2] are incorporated in the chart.

Typical Example

A storage tank with a capacity of 10,000 gal costs $2600. What would be the cost of a tank with a capacity of 18,000 gal?

The broken line between a capacity ratio of 1.8 and scale-up factor of 0.69 (for storage tanks) intersects the cost ratio at 1.5. The larger tank would then cost 1.5 × $2600, or $3900.

[1]Dickenson, R. G., *Petrol Times*, **61**, 573, 1957.
[2]Williams, R., Jr., *Chem. Eng.*, **54**, (12) 124, 1947.

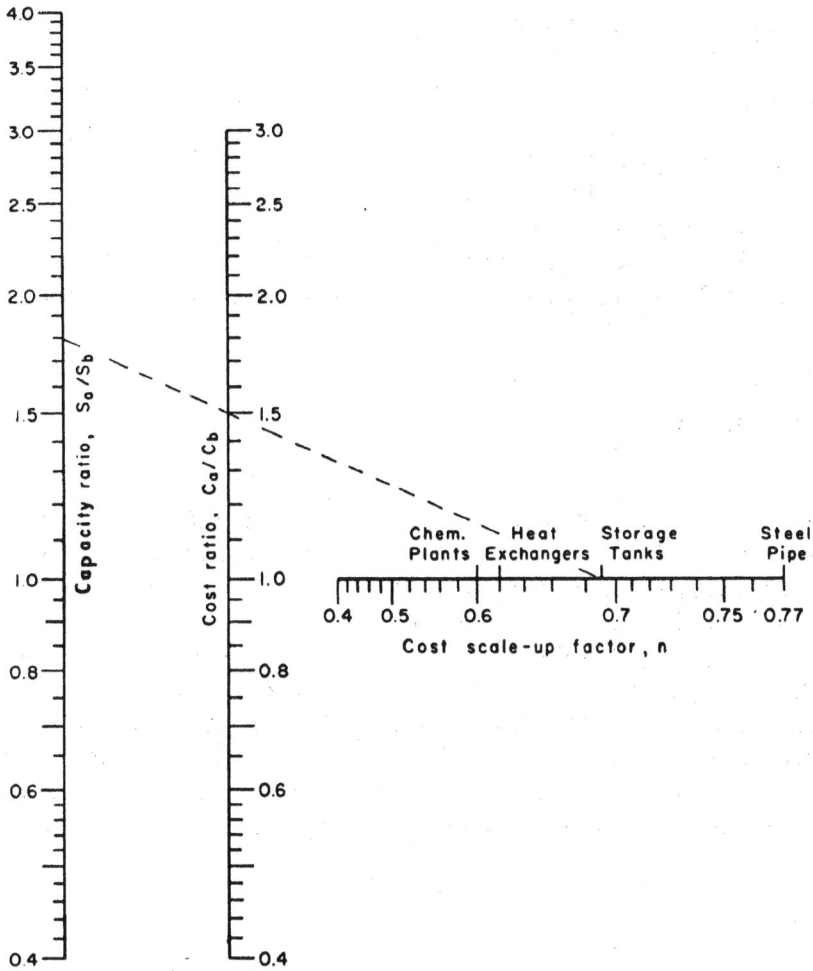

Figure 8-22

8-23   Cost of Pumping

DUKE SILVESTRINE and D. S. DAVIS

For liquids at temperatures where their viscosities are not greatly different from that of water at 68°F the cost of pumping C in cents per 1000 gallons is given by

$$C = 0.315 \frac{PH\delta}{E}$$

where P = cost of power, cents/kw-hr
H = total dynamic head, ft
δ = specific gravity
E = over-all efficiency of motor and pump, %
Figure 8-23 permits this equation to be solved readily and accurately.

Typical Example

What is the cost of pumping 1000 gal of a liquid with a specific gravity of 0.900 that exhibits a viscosity near that of water at 68°F against a total dynamic head of 300 ft if the overall efficiency of motor and pump is 50% and power costs two cents 1 kw-hr? Following the key on the chart, connect 2 on the P-scale and 300 on the H-scale with a straight line. Mark the intersection with the α-axis. Connect 50 on the E-scale and 0.900 on the δ-scale with a straight line. Mark the intersection with the β-axis. Connect the intersections on the α- and β-axes with a straight line and read the desired cost of pumping on the C-scale as 3.4 cents per 1000 gal.

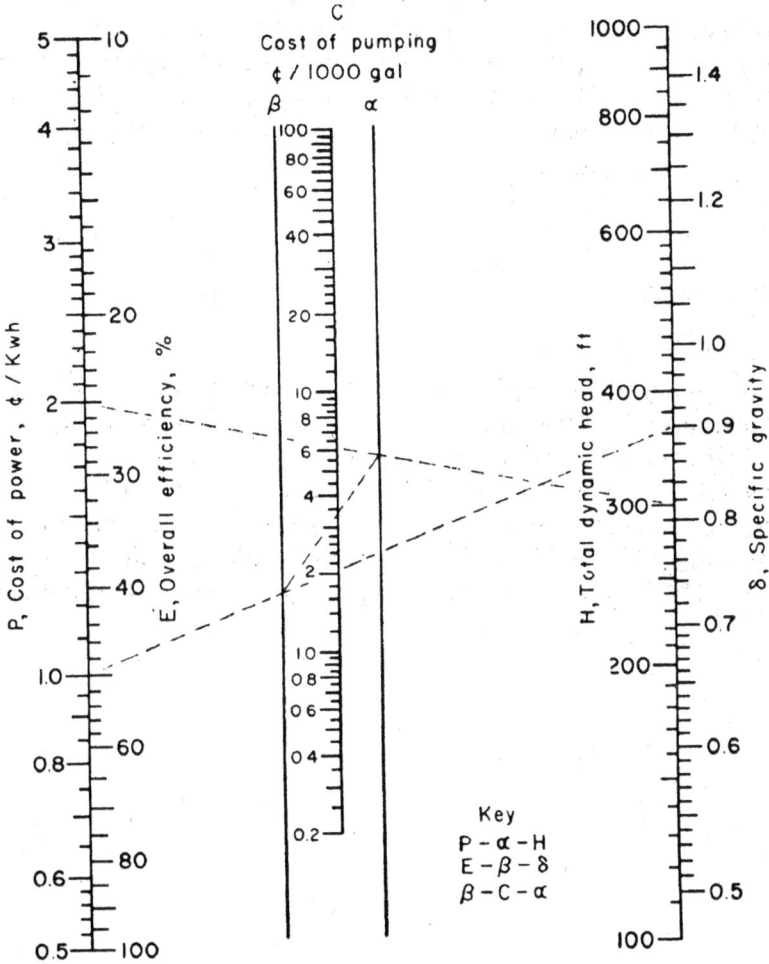

Figure 8-23

8-24   Dilution Nomograph

IRWIN COOPER

Figure 8-24 offers a convenient and rapid method of determining approximate volume of a solution of given concentration that must be added to a known volume of water to give a final solution (specific gravity approximately 1.0) of a desired concentration. As it is convenient to refer to dilute solutions in terms of ppm, the C_2-scale is graduated in this manner.

Typical Example

Approximately how many milliliters of a solution that contains 20 g of solute per 100 ml solution must be added to 1.5 liters of water to give a final solution of concentration equal to 240 ppm?

Draw a straight line extending from 20 on the C_1-scale to 1.5 on the V_1-scale. Extend it to the reference line. From this point, draw a straight line to 240 on the C_2-scale. The line intersects at 1.8 ml on V_2-scale. This is the approximate volume that will satisfy the conditions.

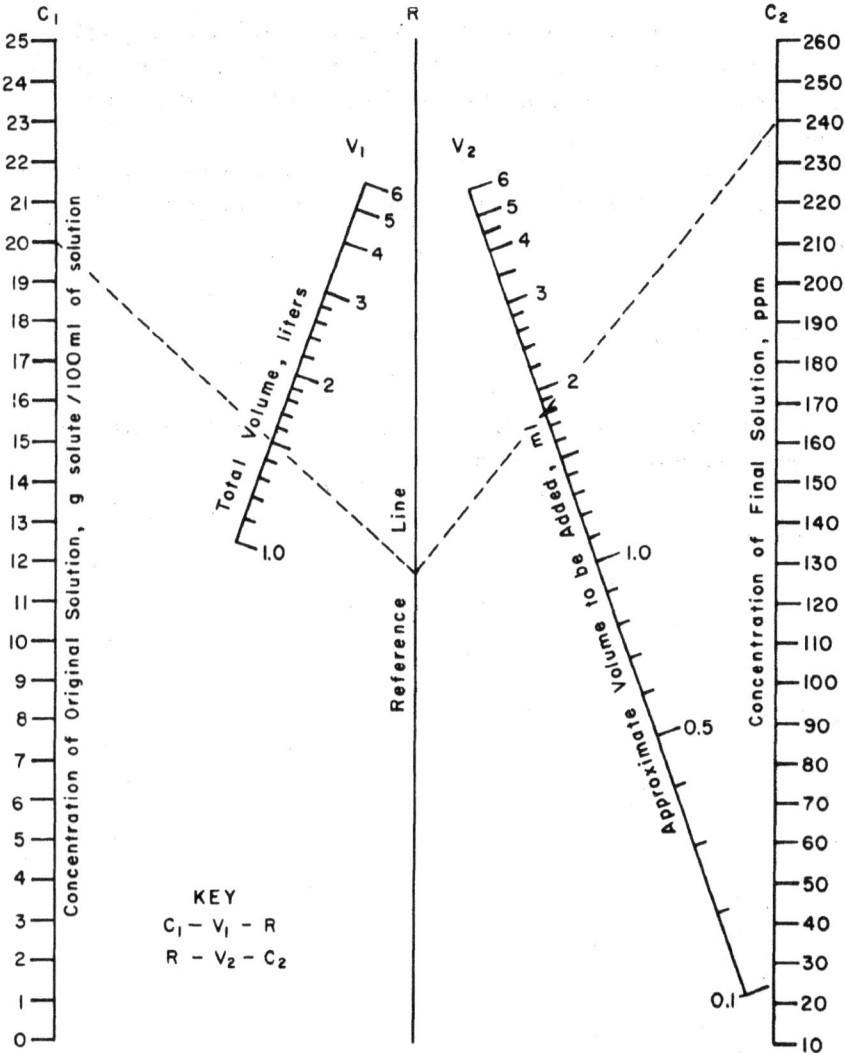

Figure 8-24

8-25   Stability of Organic Peroxides

GEORGE E. MAPSTONE

The rate of decomposition of an organic peroxide at a given temperature can be specified in terms of its half-life. This quantity represents the time required for one-half of the peroxide present to decompose.

The published half-lives of various organic peroxides in benzene solution are given as a function of temperature in Figure 8-25.

Typical Example

The dashed line shows that the half-life of benzoyl peroxide in benzene solution is 1.2 hr at 90°F.

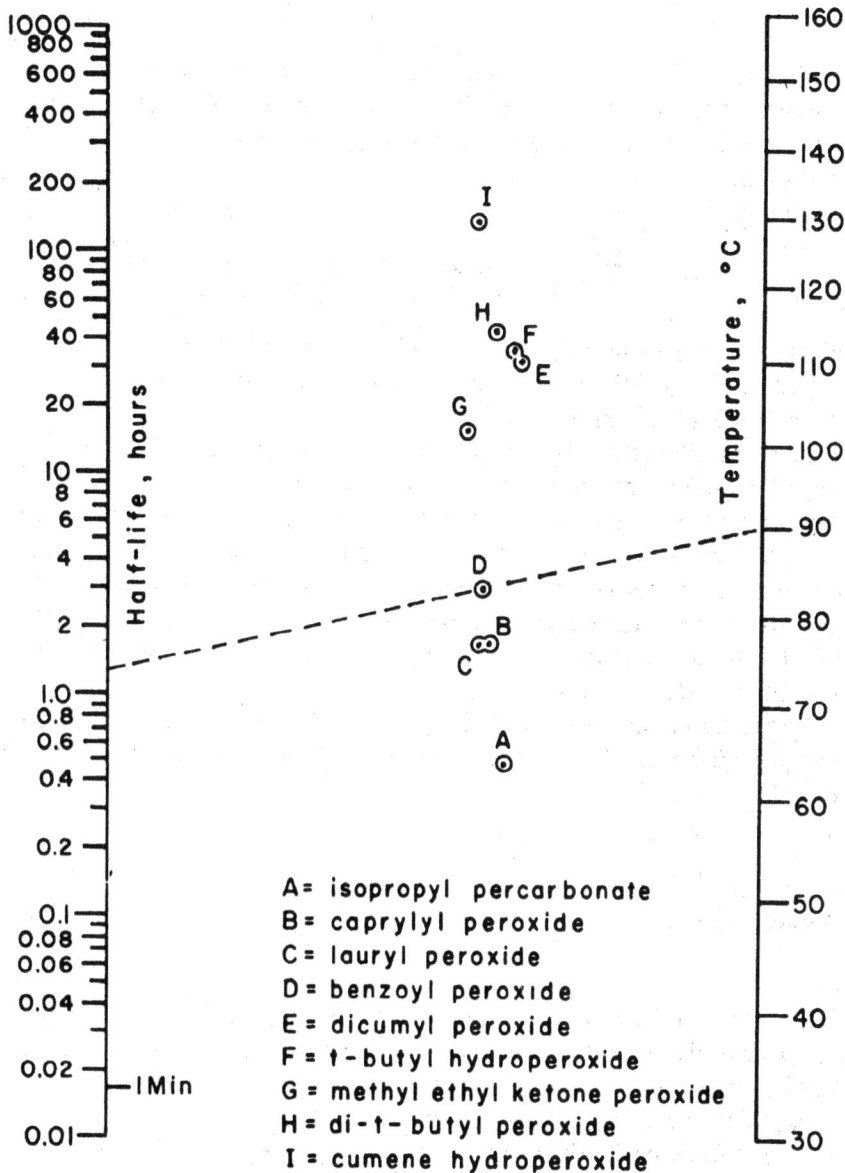

Figure 8-25

8-26   Polyurethane Reaction

M. J. CRAMER

Glycol derivatives and isocyanate derivatives react to produce polyurethane polymers. Percentage completion of reaction can be followed by titration of the unreacted isocyanate group with 0.5N hydrochloric acid.

The relation

$$100 - \frac{\text{(Batch Wt.) (Titration Vol Diff.) (N) (Equiv. Wt.)}}{10 \text{ (Sample Wt.) (Wt. of Isocyanate)}}$$

gives the percentage reaction. Total batch weight, weight of isocyanate added to the batch, equivalent weight of isocyanate, and normality N of the acid are known.

The test procedure is performed by weighing a sample, adding 25 ml of dioxane-dinormal butyl amine and 25 ml of acetone or other solvent systems, and then titrating. A blank sample is titrated also.

Figure 8-26 allows calculation for different size batches and different isocyanates of different equivalent weight. For a large series of tests on the same large batch over a period of time, a constant point will be obtained at the end of Step 3 so that the first three steps need not be repeated for each sample.

For a batch larger than 1000 g, the batch weight and isocyanate weight can be divided by the same factor to keep within range of the chart.

Typical Example

The broken line shows steps in sequence: A 350-g batch of polypropylene glycol and an isocyanate are mixed. 30 g of isocyanate are used; the equivalent molecular weight is 110. Half-normal acid is used to titrate a 5-g sample after three hours.

The difference in titration for a blank and sample is 2 ml. Percentage completion of reaction is read as 74.5%.

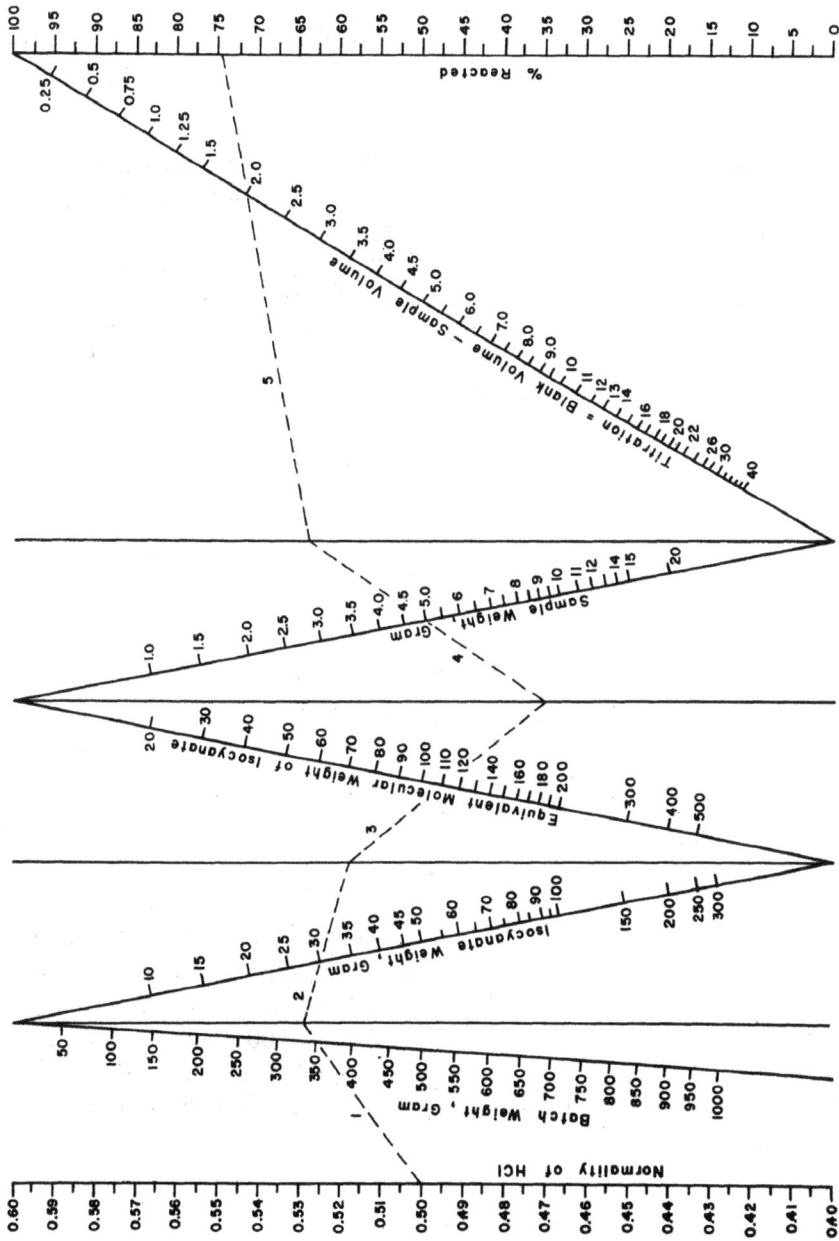

Figure 8-26

8-27   Allowable Pressure in Thick-Walled Shells

F. CAPLAN

In estimating P, the maximum allowable working pressure in lb/in.^{2} for a shell with inner and outer radii r and R inches respectively or inner and outer diameters d and D inches respectively from the allowable stress, S lb/in.^{2}, and the fractional joint efficiency E, the ASME Boiler and Pressure Vessel Code requires use of the Lamé equation

$$P = SE \left[\frac{D^2 - d^2}{D^2 + d^2} \right]$$
$$= SE \left[\frac{R^2 - r^2}{R^2 + r^2} \right]$$

or

$$\left[\frac{D}{d} \right]^2 = \left[\frac{R}{r} \right]^2 = \frac{SE + P}{SE - P}$$

when the thickness of the shell exceeds $\frac{1}{2}$ of the inner radius. This equation can be solved by means of Figure 8-27.

Typical Example

Use of the chart is illustrated as follows: What is the maximum allowable working pressure for a shell with inner and outer radii of 2.89 and 5.00 in. if the product SE of allowable stress and fractional joint efficiency is 20,000 lb/in.^{2}? Align R = 5.00 with r = 2.89 and mark the intersection with the α-scale; align this intersection with SE = 20,000, and read the maximum allowable working pressure on the P-scale as 10,000 lb/in.^{2}

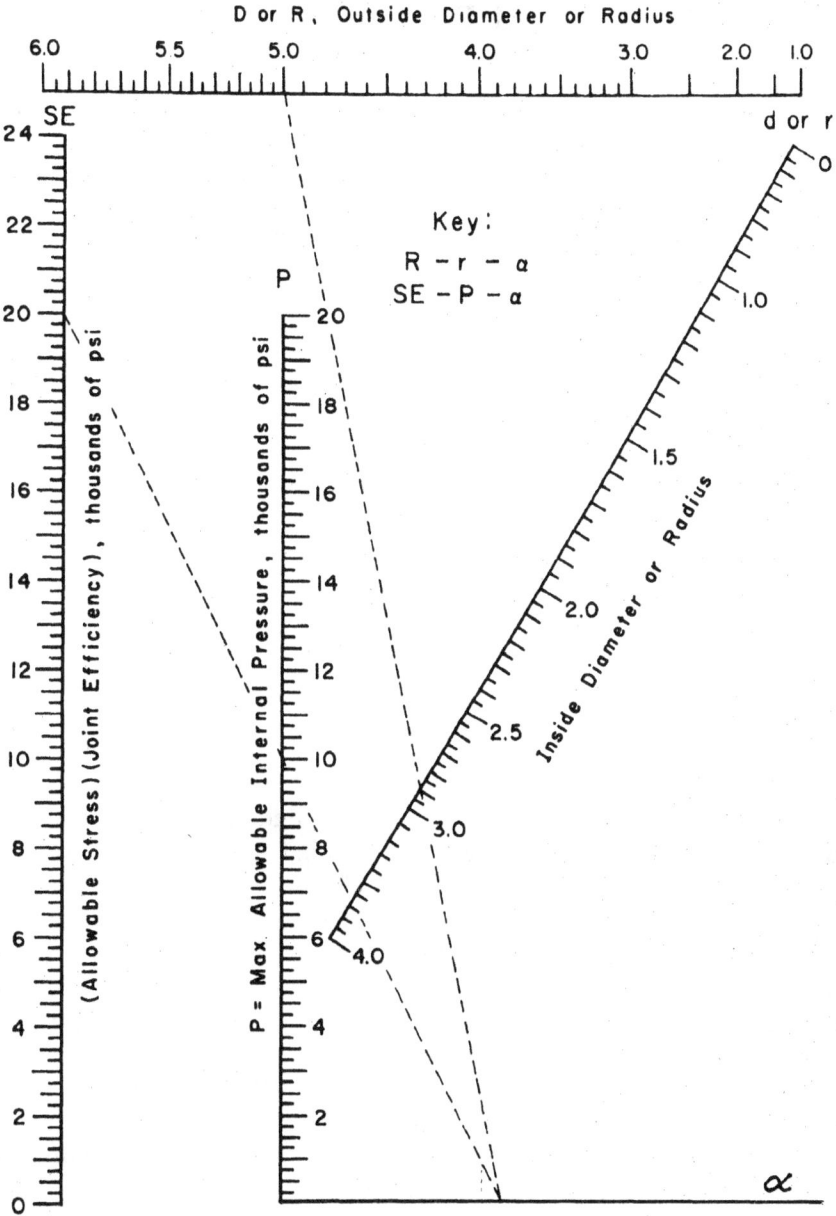

Figure 8-27

**8-28   Thicknesses of Cylindrical Shells for
                    Unfired Pressure Vessels**

ELIZABETH D. SHROFF

Figure 8-28 presents a simplified method for determining the required plate thickness for cylindrical pressure vessels to be designed to conform to the ASME Boiler & Pressure Vessel Code, Section VIII — "Unfired Pressure Vessels" (1959 edition). The nomograph is based on the equation

$$t = \frac{PR}{Se - 0.6\,P}$$

where   t   = minimum required thickness of shell plates exclusive of corrosion allowance, in.

   P   = design pressure, $lb/in.^2$

   R   = inside radius of shell before corrosion allowance is added, in.

   S   = max allowable stress value, $lb/in.^2$

   e   = efficiency of longitudinal joint, %

Typical Example

Determine the thickness required for a cylindrical-shell with an ID of 96 in. and design pressure of 100 $lb/in.^2$ Maximum allowable stress value is 15,000 $lb/in.^2$ and joint efficiency is 95%.

Join 48 on the R-scale with 100 on the P-scale to intersect the t_1-scale. From this point join 15,000 $lb/in.^2$ on the S-scale to intersect the t_2-scale. Now join this intersection with 95 on the e-scale and read the value on the t_1-scale between 5/16 in. and 3/8 in. Use the larger value of 3/8 in. for shell thickness.

If the stress value were held constant at 13,750 $lb/in.^2$ and the joint efficiency at 80%, then the value of thickness t may be read directly at the intersection of the t_1-scale by the line joining R-scale with the P-scale. In above example for R = 48 in. and P = 100 $lb/in.^2$, required value of t = 7/16 in. is obtained on the t_1-scale for fixed stress value of 13,750 and joint efficiency of 80%.

The t_2-scale gives required value for variable values of R, P and S and a fixed value of 80% for joint efficiency. In above example for

R = 48 in., P = 100 lb/in.^{2} and S = 15,000 lb/in.^{2}, value of t on t$_2$-scale between 3/8 in. and 7/16 in. is obtained for a joint-efficiency value of 80%.

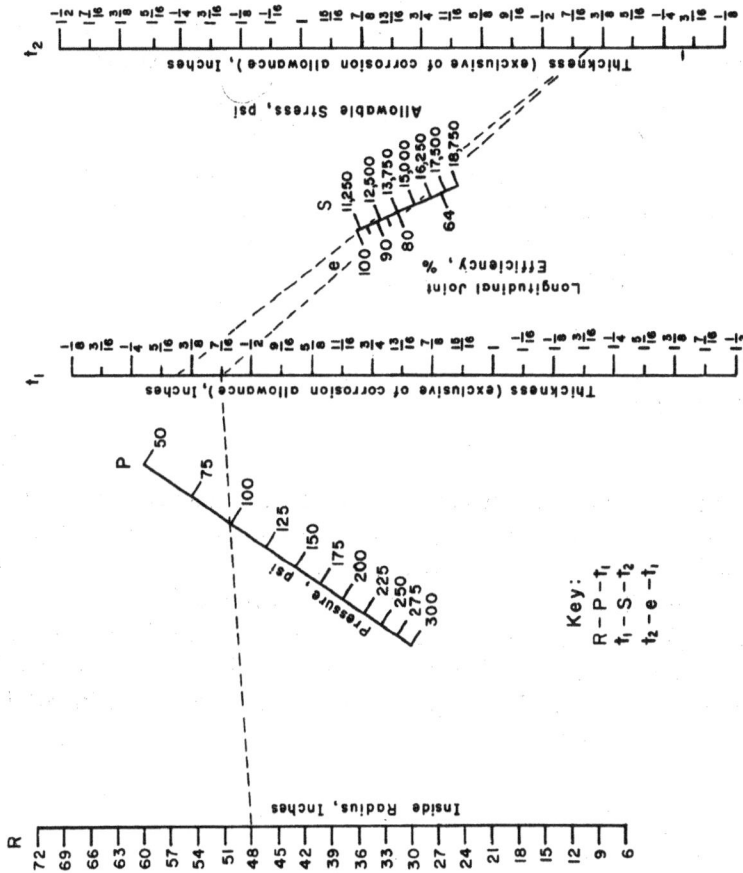

Figure 8-28

8-29   Gaseous Nitrogen for Purging

VINCENT J. NAIMOLI and D. S. DAVIS

Inert characteristics of gaseous nitrogen make it an invaluable purge medium. The gas usually is drawn from a large cascade or reservoir into a "running tank." The "run" tank then is pressurized and its contents are expanded as desired for purging. Total consumption of nitrogen can be computed by means of the formula

$$V_2 = 0.00909 \, (p_1 + 14.7) \, V_1 N$$

where V_2 = total consumption of nitrogen, standard ft^3
p_1 = purge pressure at tank, lb/in.^{2} gage
V_1 = volume of "run" tank, gal
N = number of pressurizations

Typical Example

How many standard cubic feet of nitrogen are consumed in purging rocket-fuel lines over nine runs? Purge gas flows from a two-gal tank that is refilled from a nitrogen cascade after each run. Purge pressure is 200 lb/in.^{2} gage. On Figure 8-29 connect 2 on the V_1-scale and 200 on the p_1-scale with a straight line, marking the intersection with the α-axis. Connect this intersection and 9 on the N-scale with a straight line and read the total volume of nitrogen required for purging as 35 standard ft^3 on the V_2-scale.

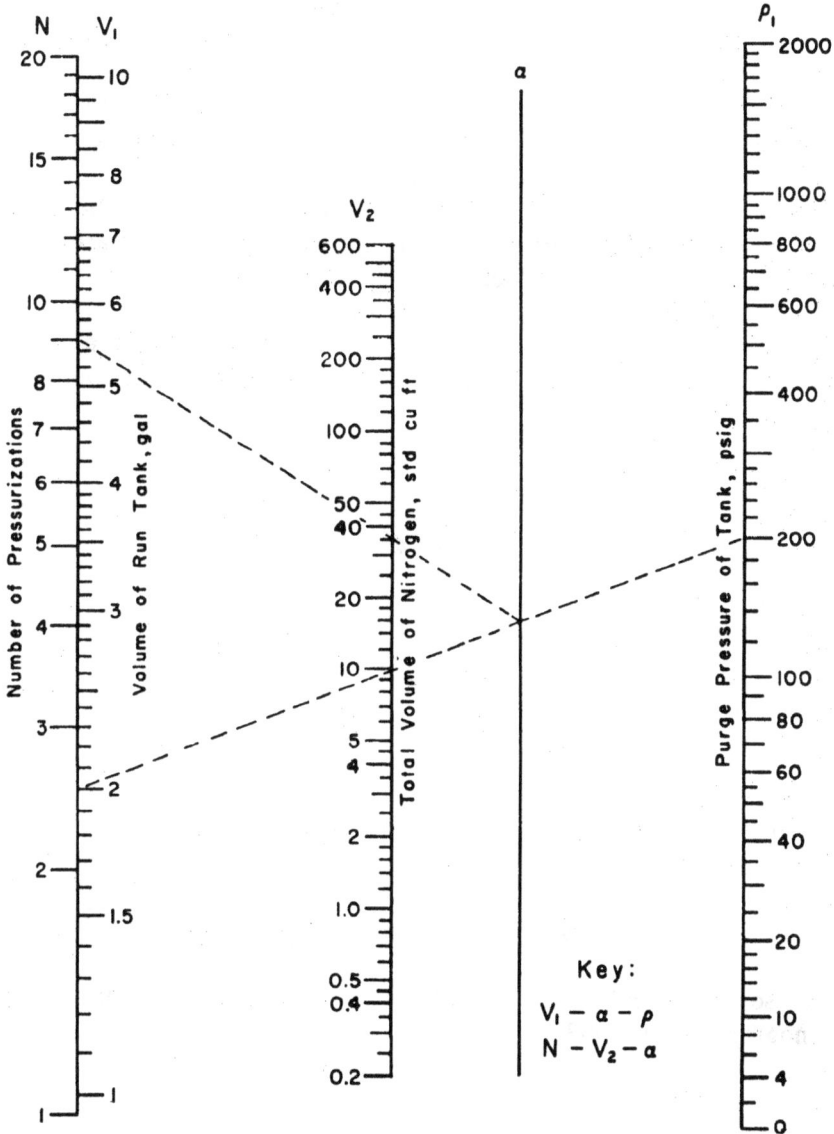

Figure 8-29

8-30   Purge Calculations

K. A. HARPER

Many cases arise in chemical processing where an inventory of material is purged by adding small increments of a new material and withdrawing small increments of the mixture. Calculating the time required (or material required) to purge to a given percentage of the equilibrium change is often necessary.

If it is assumed that the material withdrawn is always of the same composition as the inventory, this calculation is straight-forward.

Let P = new material added, % of inventory

   C = % of new material present in inventory

A differential equation representing the effect of adding a small increment of new material and withdrawing the same amount of (perfectly mixed) inventory is as follows:

$$dC = \frac{dP\,(100-C)}{100}$$

$$dP = \frac{100\,dC}{100-C}$$

Integrating

$$P = 100 \ln \frac{100}{100-C}$$

$$= 230.3 \log \frac{100}{100-C}$$

Let θ = days elapsed

   M = average daily makeup, % of inventory

Then

$$\theta = \frac{P}{M} = \frac{230.3}{M} \log \frac{100}{100-C}$$

These equations have been converted into the accompanying nomograph, Figure 8-30.

Typical Example

Occasionally an FCC unit is operating on one type of catalyst, and a change to a new type of catalyst is accomplished by simply using the new catalyst for makeup. This type of problem prompted development of this equation initially. Actually, the stack losses are not the same concentration of the new catalyst as the unit inventory. However, this method is still useful in approximating the change-over time. Even chemical analysis is subject to some difficulty because of fluctuations of the content of key component in the new catalyst.

For example, calculate the time required to reach 90% of new catalyst in the unit if makeup is 2% of inventory daily. Connecting 2 on the M-scale to 90% on the C-scale gives the time required as 115 days.

The nomograph can easily be used to calculate the time to get from one concentration to another. At 2% makeup, the time to get from 60% to 90% would be 115 minus 46 or 69 days.

Other Applications

A less obvious use of Figure 8-30 is in changes of contaminant levels of the catalyst. Assume the contaminants added daily as being deposited on the new catalyst exclusively and calculate results as before.

Other applications include accumulators, storage tanks, and tank cars, which will purge similarly if continuous mixing occurs. This may be from off-specification to acceptable products or from one specification to another.

The last scale on the left is total purge (P) as percentage of inventory. For some problems it will be simpler to use this scale. In that case, C can be read directly on the adjoining scale. For example, addition of makeup amounting to 100% of inventory gives 62% of the new material in the inventory.

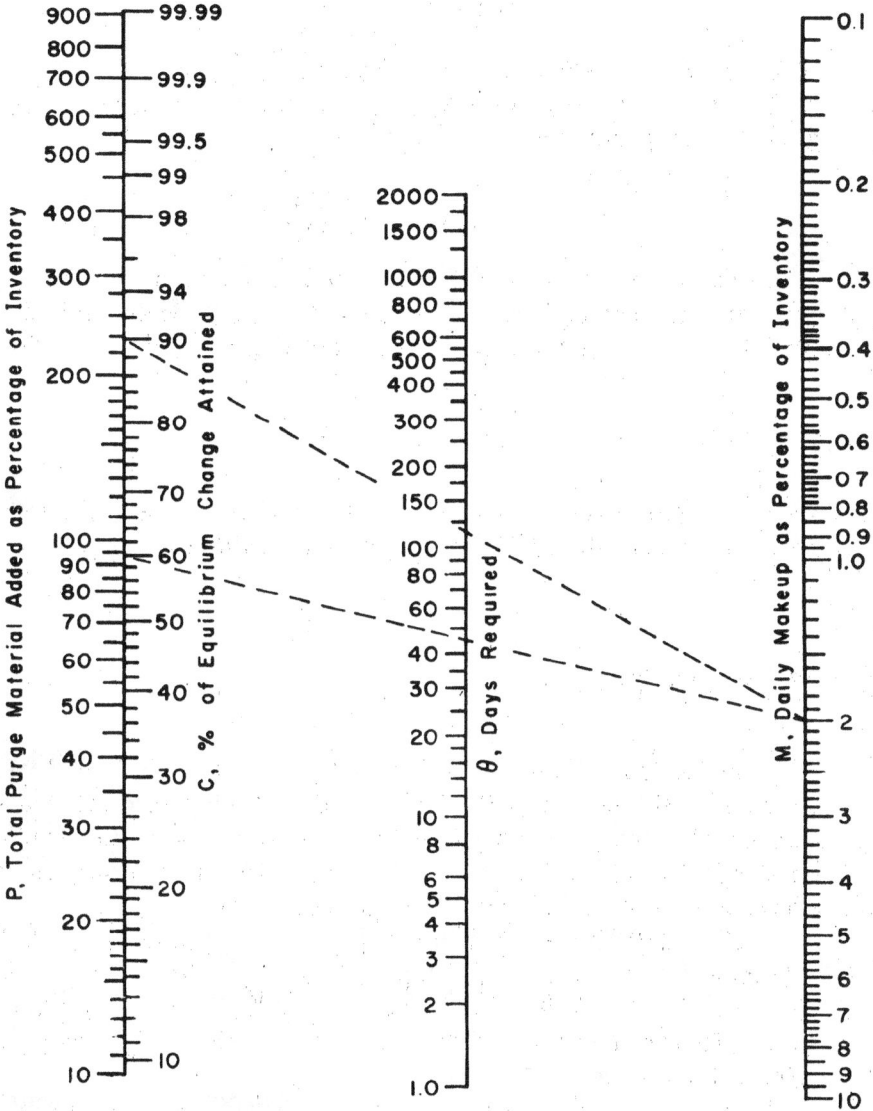

Figure 8-30

8-31   Theoretical and Typical W/Q Ratios for Cryogenic Refrigeration Systems

PAUL E. LOVEDAY

The ratio of theoretical work required W to heat Q, absorbed isothermally at a lower level T_2 (and rejected isothermally at T_1), is given by the well-known expression

$$\left(\frac{W}{Q}\right)_{Theo} = \frac{T_1 - T_2}{T_2}$$

Because of electrical and mechanical losses and heat leak, Δt's, and Δp's in heat exchangers, lines, and other process equipment, typical values of W/Q are significantly higher and can be expressed as

$$\left(\frac{W}{Q}\right)_{Typ} = \frac{\left(\dfrac{W}{Q}\right)_{Theo}}{E}$$

where E is the over-all efficiency of the process. The table below provides typical values of E for various refrigeration systems.

Typical Examples

1. For a liquid helium system with $T_2 = 4.2°K$ and $T_1 = 300°K$, determine $(W/Q)_{Theo}$ and $(W/Q)_{Typ}$ if $E = 0.1$. Use Figure 8-31. Connect $(T_1 - T_2) = 295.8°K$ with 4.2 on the T_2-scale and read 70 on the $(W/Q)_{Theo}$ scale. Then connect this point with 0.1 on the efficiency scale and read 700 on the $(W/Q)_{Typ}$ scale.

2. For a fluoromethane system with $T_2 = 258°K$ and $T_1 = 300°K$, determine $(W/Q)_{Theo}$ and $(W/Q)_{Typ}$ if $E = 0.45$. Connect $(T_1 - T_2) = 42°K$ with 258 on the T_2-scale, and read 0.16 on the $(W/Q)_{Theo}$ scale. Connect this point with 0.45 on the efficiency scale and read 0.36 on the $(W/Q)_{Typ}$ scale.

These examples show the significant variations in power requirements for refrigeration systems as a function of the T_2 operating level.

Refrigeration system	Overall efficiency, E
Fluoromethanes	
(Freon 12 & 22, Generon 12 & 22,	
Ucon 12 & 22)	0.4 -0.5
Helium	0.1
Nitrogen	0.25-0.35
Hydrogen	0.20-0.30
Argon	0.25-0.35
Oxygen*	0.25-0.35

* Oxygen included for calculations involving air-separation plants.

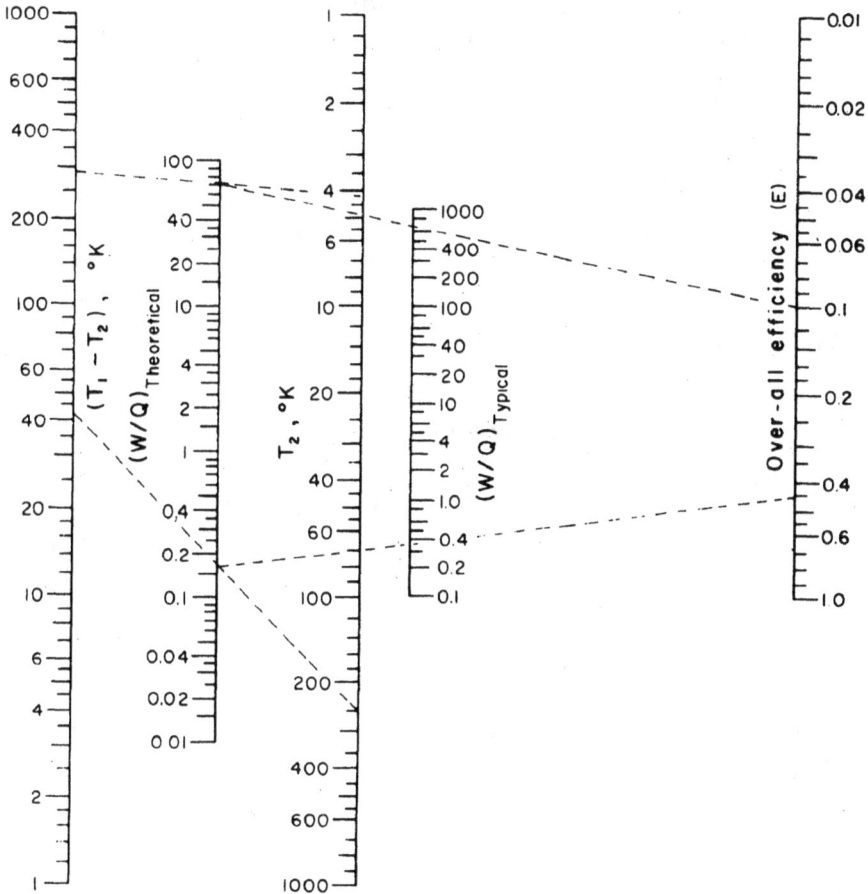

Figure 8-31

8-32   Resin Prices on a Volume Basis

T. J. GRAIL

In many molding applications, the choice of resin to be used may be narrowed by specifications to two or three possibilities. Economics then becomes the determining factor.

Although raw-materials costs are reported on a weight basis, the final product is of fixed dimensions. Resin prices should therefore be compared on a volume, rather than a weight, basis.

Figure 8-32 relates weight price ($/lb) to volume price (¢ /in.³) and to density (g/cm³). It is based on the following equation:

$$V = 3.61 \, WD$$

where V = volume price
          W = weight price
          D = density

Typical Example

A resin having a density of 1.42 g/cm³ is selling for $0.29/lb. What is its price on a volume basis? Connecting 1.42 on the D-scale with 0.29 on the W-scale gives a value on the V-scale of 1.5 ¢ /in.³

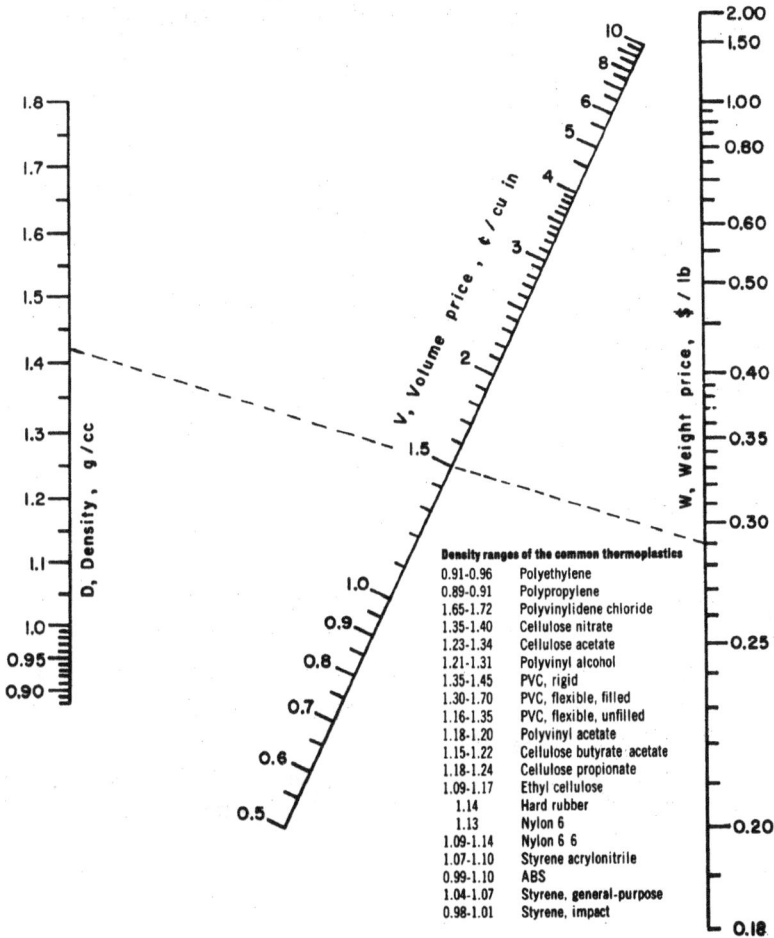

Density ranges of the common thermoplastics

0.91-0.96	Polyethylene
0.89-0.91	Polypropylene
1.65-1.72	Polyvinylidene chloride
1.35-1.40	Cellulose nitrate
1.23-1.34	Cellulose acetate
1.21-1.31	Polyvinyl alcohol
1.35-1.45	PVC, rigid
1.30-1.70	PVC, flexible, filled
1.16-1.35	PVC, flexible, unfilled
1.18-1.20	Polyvinyl acetate
1.15-1.22	Cellulose butyrate acetate
1.18-1.24	Cellulose propionate
1.09-1.17	Ethyl cellulose
1.14	Hard rubber
1.13	Nylon 6
1.09-1.14	Nylon 6 6
1.07-1.10	Styrene acrylonitrile
0.99-1.10	ABS
1.04-1.07	Styrene, general-purpose
0.98-1.01	Styrene, impact

Figure 8-32

8-33   Screen Characteristics

B. B. KLIMA and D. S. DAVIS

When using screens, chemical engineers compute opening sizes and percentages of open area through use of the equations

$$N(D+M)=1000$$

and

$$A=(1000-ND)^2 \times 10^{-4}$$

where   N = number of meshes per linear inch
   D = diameter of wire, 0.001 in.
   M = mesh opening between wires, 0.001 in.
   A = percentage of open area
   Figures 8-33a, 8-33b, and 8-33c facilitate these computations.

Typical Examples

The broken index line on Figure 8-33a shows that the mesh opening for a 5-mesh screen is 0.156 in. when the diameter of each wire is 0.044 in.   The broken index line on Figure 8-33b, for finer screens, shows that the mesh opening for a 14-mesh screen is 0.046 in. when the diameter of each wire is 0.025 in.

The index line on Figure 8-33c shows that the percentage of open area for a 60-mesh screen is 34 when the diameter of each wire is 0.007 in.

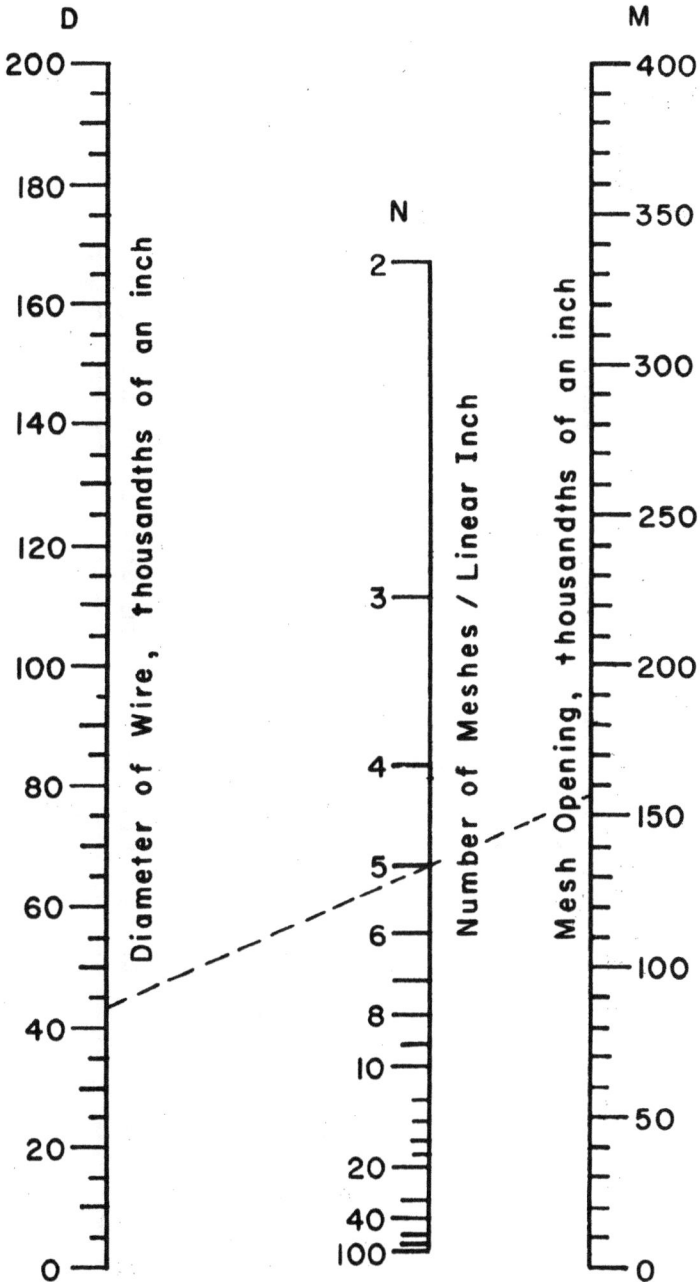

Figure 8-33a

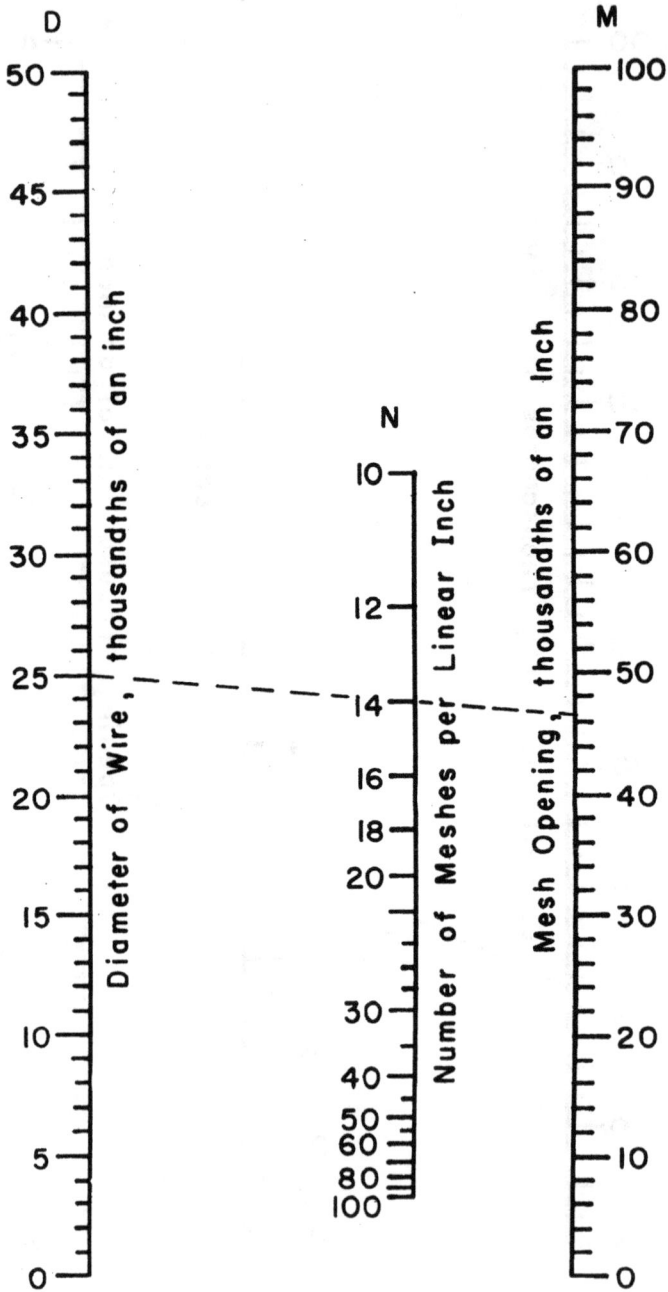

Figure 8-33b

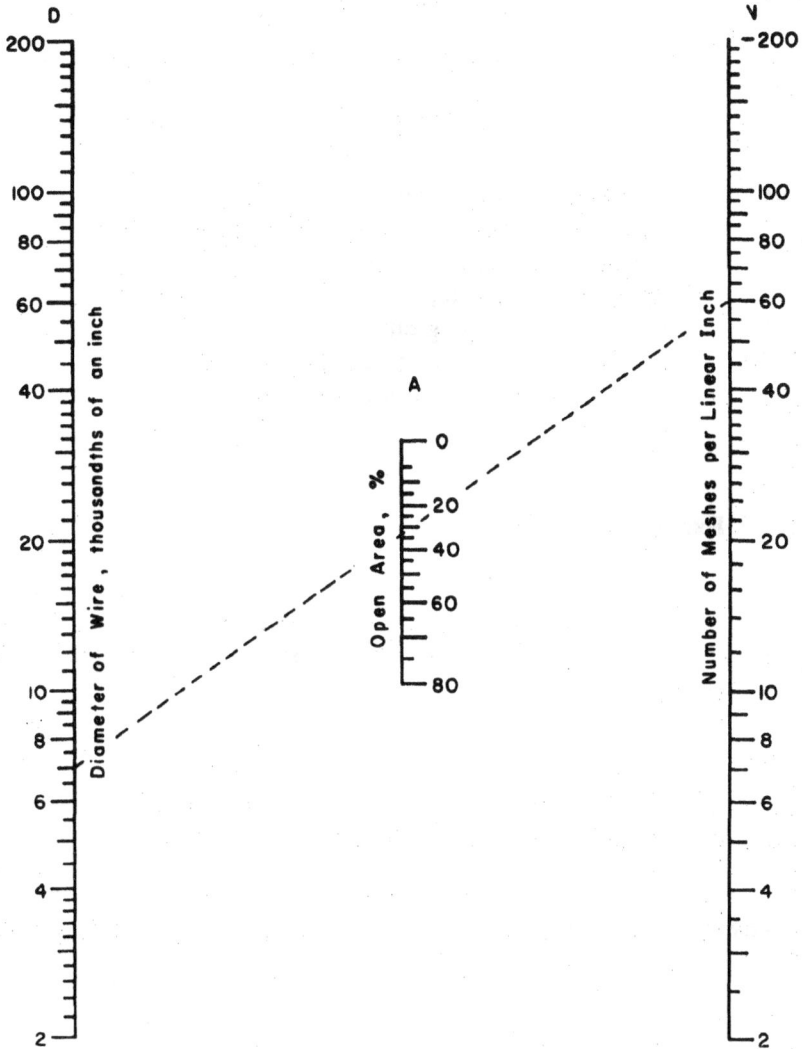

Figure 8-33c

8-34   Sedimentation Rates

DENNIS E. DRAYER

Stoke's formula[1] relating the settling rates of spherical grains is

$$u = \frac{gD^2 (\rho_s - \rho)}{18\,\eta}$$

where u = terminal velocity, cm/sec
       g = acceleration due to gravity, 981 cm/sec^2
       D = diameter of sphere, cm
       ρ_s = density of sphere, g/cm^3
       ρ = density of fluid, g/cm^3
       η = coefficient of viscosity of fluid, poises
Figure 8-34 permits rapid solution of the equation.

Typical Example

What is the terminal settling velocity for 0.01 cm diameter spheres of density 2.0 g/cm^3 in water at 20°C? For water at 20°C, $\rho = 1.0$ and $\eta = 0.01$.

$$\rho_s - \rho = 2.0 - 1.0 = 1.0$$

Connect D = 0.01 with $\eta = 0.01$. Then connect the reference line intersection with $\rho_s - \rho = 1.0$ and read u = 0.55 cm/sec.

[1]Perry, J. H. *Chemical Engineers' Handbook*, 3rd ed, p. 937, McGraw-Hill Book Co., New York (1950).

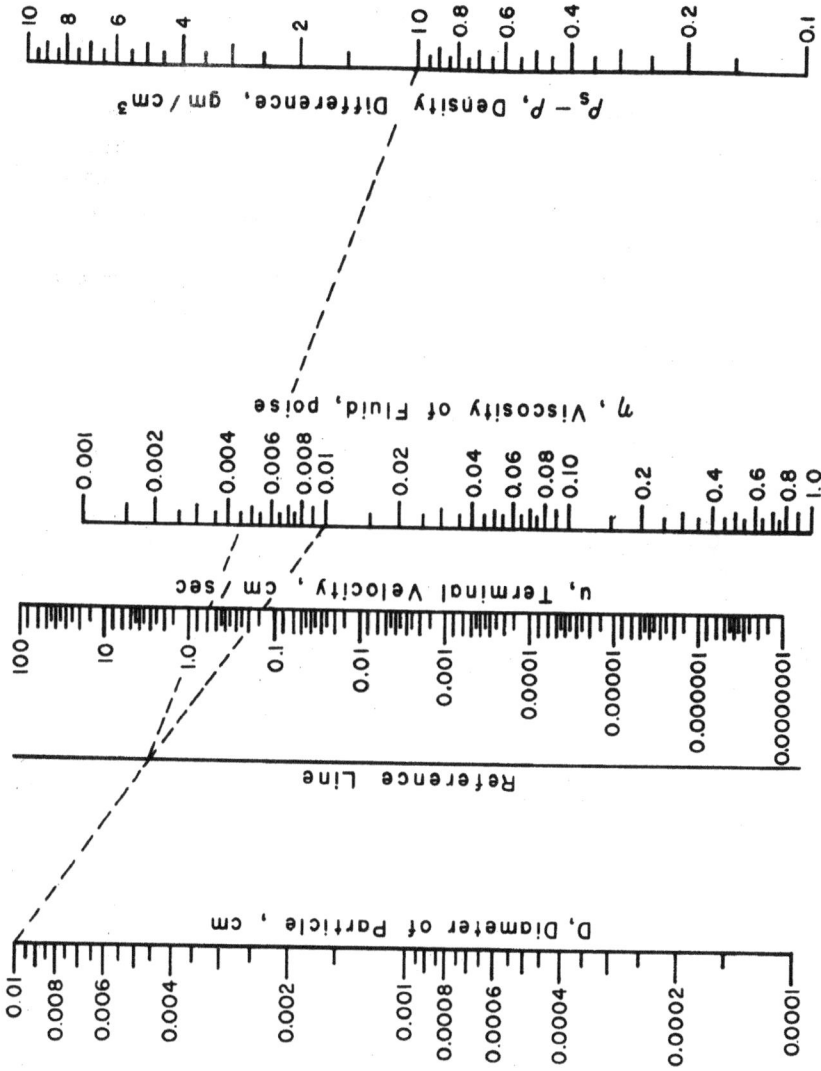

Figure 8-34

8-35   Speed vs Sheave Sizes

FRED RUPPERT

Use Figure 8-35 to determine quickly the sheave size required to operate a V-belt-driven piece of processing equipment when the motor speed is 3450 rpm. Place a straightedge on the driven shaft scale at the desired speed and connect to the motor shaft sheave size scale at any standard sheave size that is available. Read the other sheave size required directly from the intersection of the driven shaft sheave size scale and the straightedge. Correct values for driven-shaft rpm may also be found.

Typical Example

A straightedge connected to a 6 in. sheave size on motor shaft sheave size scale and intersecting 3 in. on the driven shaft sheave size scale indicates a speed of 6900 rpm.

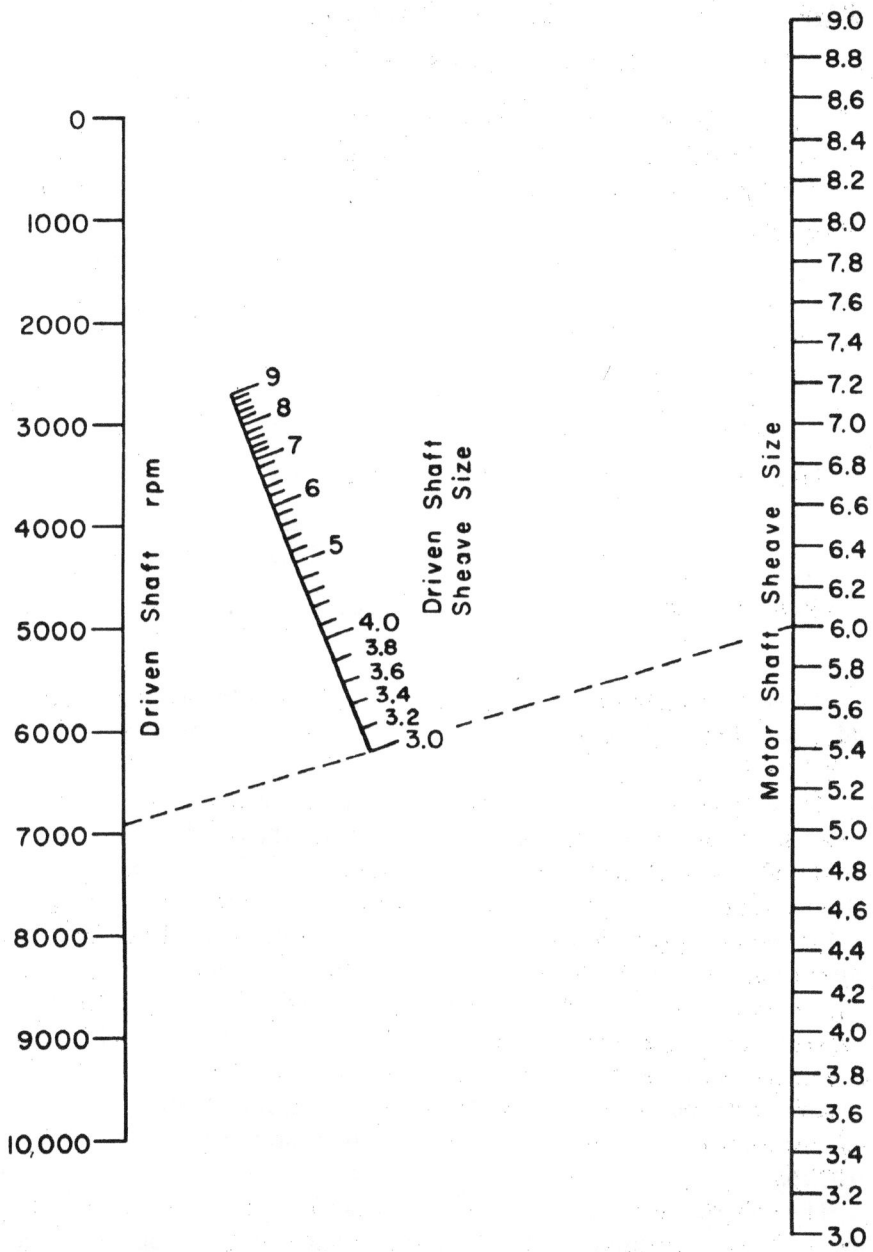

Figure 8-35

8-36   Solution Blending

D. W. WILLISTON and D. S. DAVIS

An amount of solution weighing A lb with a concentration of a
% is often mixed with B lb of solution having concentration of b %
to produce a blend that contains m % — in accordance with the
balance

$$Aa + Bb = (A + B) m \qquad (1)$$

written

$$\frac{A}{B} = \frac{m-b}{a-m} \qquad (2)$$

This equation is the basis of Figure 8-36. ·

Typical Example

1) How much 20% acid is needed to dilute 60 lb of 80% acid to
44%? That is, when A = 60 lb, a = 80%, b = 20% and m = 44%,
what is B?
Connect $44-20 = 24$ on the $(m-b)$-scale and $80-44 = 36$ on
the $(a-m)$-scale with a straight line: Connect the a-scale intersection
with 60 on the A-scale and read the amount of 20% acid as 90 lb on
the B-scale. If A were 6 or 600, B would be 9 or 900, respectively.
2) What quantities of 22% and 78% alcohol should be mixed to
yield 100 lb of 40% alcohol? Connect $40-22 = 18$ on the $(m-b)$-
scale and $78-40 = 38$ on the $(a-m)$-scale with a straight line and
mark the intersection with the a-scale. A trial with A = 30 (not
indicated on the chart) yields B = 63.3, so that the sum is 93.3 —
short of the necessary 100 lb. As shown, increasing A to 32 increases
B to approximately 68, so that the sum is approximately the desired
100 lb.
The chart can be used for blending (a) with pure solvent (for which
b = 0), (b) by volume when the solutions have nearly equal densities,
(c) gases at the same temperature and pressure, and (d) granular solids
and semisolids.

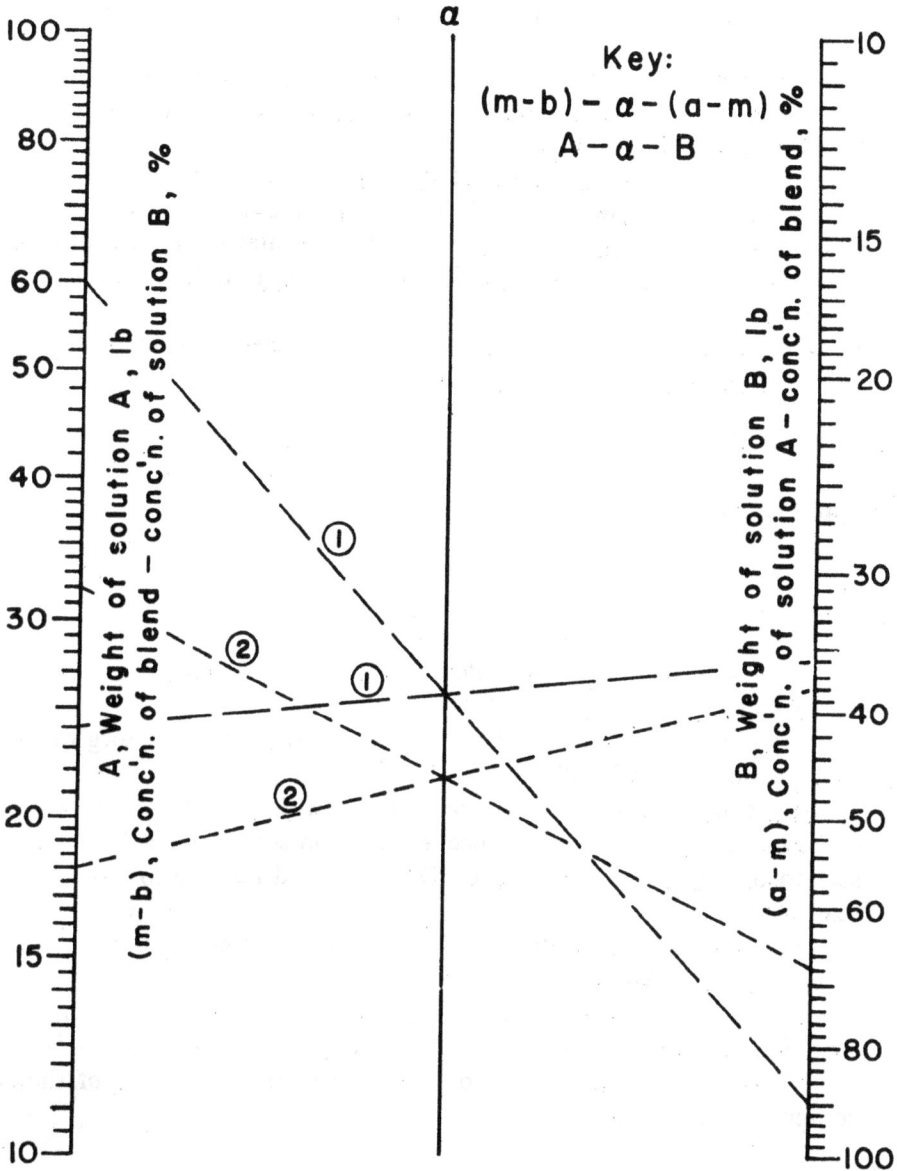

Figure 8-36

8-37   Nomograph for Mixing Solutions

R. J. LAMBERT

Figure 8-37 can be used to determine quickly the wet weight of a solution to be added to another when the wet weight to be added is dependent upon a definite solids ratio.

In the case of the nomograph, the solids ratio is equal to 1.00. If a different solids ratio is desired, scales B and D should be shifted. For example, if a solids ratio of 0.6 is desired, scales would be shifted so that 1667 lb on scale B and 166.7 lb on scale D fall on the index line (top).

In this manner a nomograph is simply constructed to determine the wet weight of a solution to be added to another for any solids ratio.

Typical Example

Given the following factors:
    Solids of solution A = 24.0%
    Wet weight solution A added to mixer = 1500 lb
    Solids of solution B = 22.0%
find the wet weight of solution B to be added to solution A to give a solids ratio (A/B) of 1.00.

Draw a line connecting 24% on scale A with 1500 lb on scale F (Line 1). Now draw a line connecting 22% on scale E with the intersection of Line 1 on scales C and D and extend the line to intersect scale B.

This point of intersection on scale B gives wet weight of solution B to be added as 1640 lb.

Scales C and D can be used to give the total or individual dry solids weight for each solution. In the above example this is 720 lb dry solids for the finished solution or 360 lb dry solids for each of the component solutions.

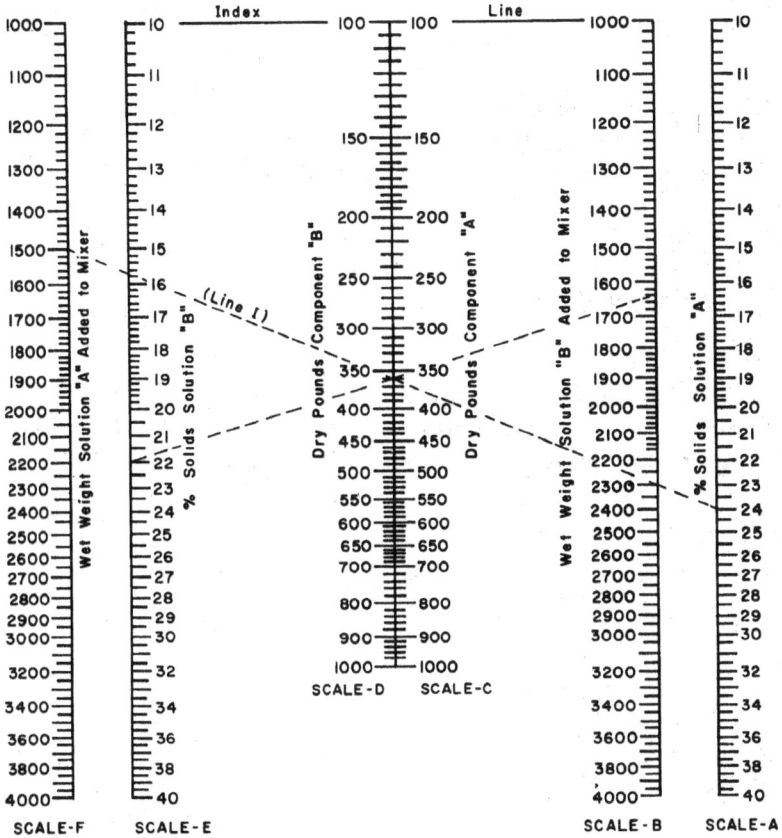

Figure 8-37

8-38   Boiling Temperatures of Sulfuric Acid

R. F. BATTEY

The line-coordinate chart of Figure 8-38 is based on the integrated Clausius-Clapeyron equation relating vapor pressure and temperature:

$$\log P = A - B/T$$

where P = vapor pressure, mm Hg
       T = absolute temperature, °K
       A, B are functions of H_2SO_4 concentration.

Typical Example

Find the boiling temperature of a 65% H_2SO_4 solution at 5 in. Hg absolute pressure. Connecting 5 in. on the right-hand scale and 65% on the middle scale, read 216°F on the left-hand scale.

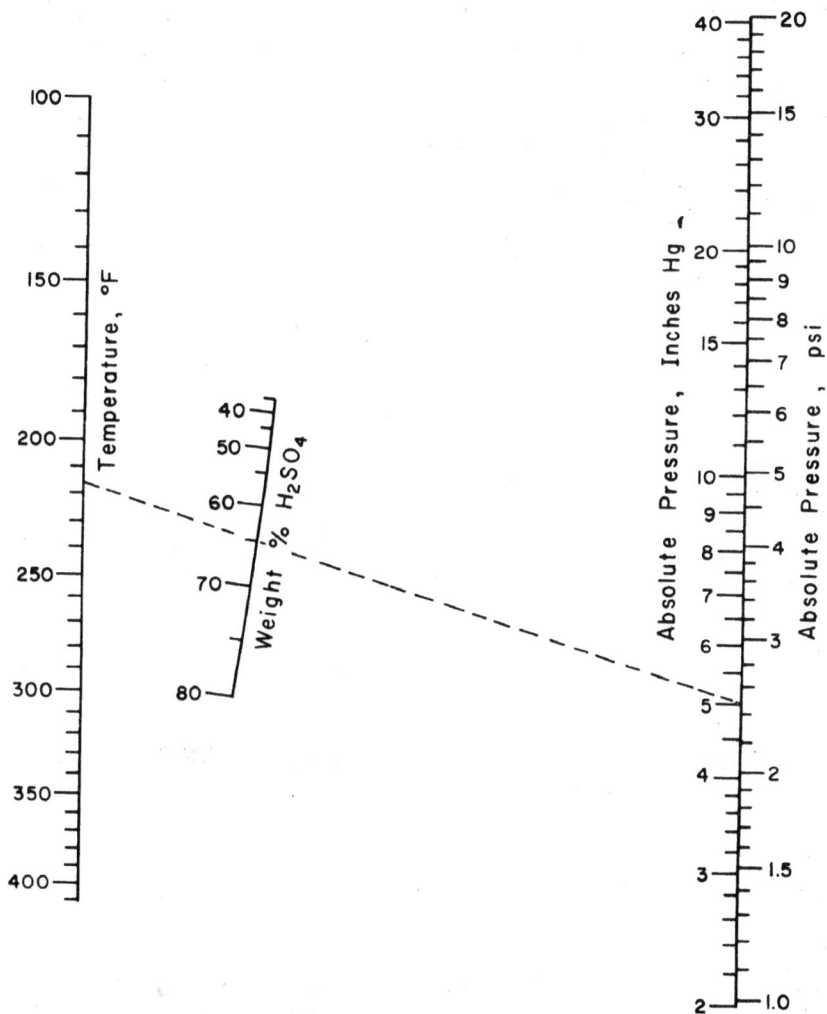

Figure 8-38

8-39   Temperatures for Doubled Reaction Rates

C. R. NODDINGS, G. M. MULLET and D. S. DAVIS

Figure 8-39 permits rapid determination of the operating temperature T_2°K, which must be attained in order to double the rate of a reaction over its value at a given T_1. The corresponding activation energy must be known.

Activation energy E in cal/g-mole, is related to reaction rates k_1 and k_2 at corresponding temperatures T_1 and T_2 (°K) by the equation

$$\frac{k_2}{k_1} = 10^{-n} \tag{1}$$

where

$$n = \frac{E\left(\frac{1}{T_2} - \frac{1}{T_1}\right)}{4.574} \tag{2}$$

When reaction rates are doubled,

$$\frac{k_2}{k_1} = 2 \tag{3}$$

and equations (1) and (2) can be reduced to

$$\frac{1}{T_2} = \frac{1}{T_1} - \frac{1.3769}{E} \tag{4}$$

This latter equation can be readily solved by use of the nomograph.

Typical Example

When the activation energy is 19,000 cal/g-mole, at what temperature must the reaction be carried out so that the reaction rate is twice that at 520°K?

Connect 19,000 on the activation-energy scale and 520 on the T_1-scale with a straight line. Read the desired temperature as 540°K on the T_2-scale of the same set.

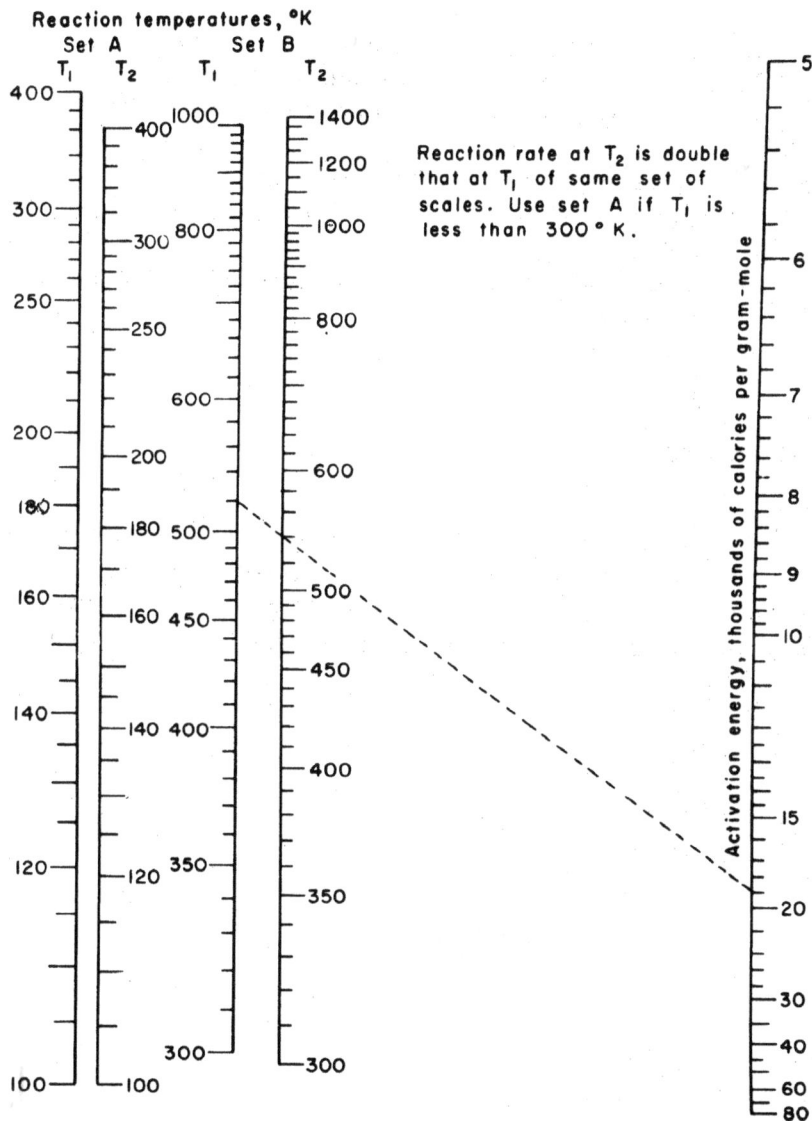

Reaction temperatures, °K

Reaction rate at T₂ is double that at T₁ of same set of scales. Use set A if T₁ is less than 300°K.

Activation energy, thousands of calories per gram-mole

Figure 8-39

8-40   Volumes of Standard Dished Heads
at Various Depths

J. F. KUONG

In estimating volumes of product heels for purposes of inventory, the chemical engineer needs to compute volumes of standard dished heads for vertical tanks at various depths by means of following formula:

$$V = 0.01363 \ H^2L - 0.004545 \ H^3$$

where V = volume of liquid in dish, gal (excluding flanged portion)

      H = depth of liquid in dish, in.

      L = radius of dish, in. (usually equal to diameter of tank minus 6 in.)

Typical Example

This computation can be performed readily by means of Figure 8-40, on which the broken index line shows that 190 gal of liquid are contained in the standard dished head of a vertical tank, when the depth of the liquid is 10.5 in. and the radius of the dished head is 130 in.

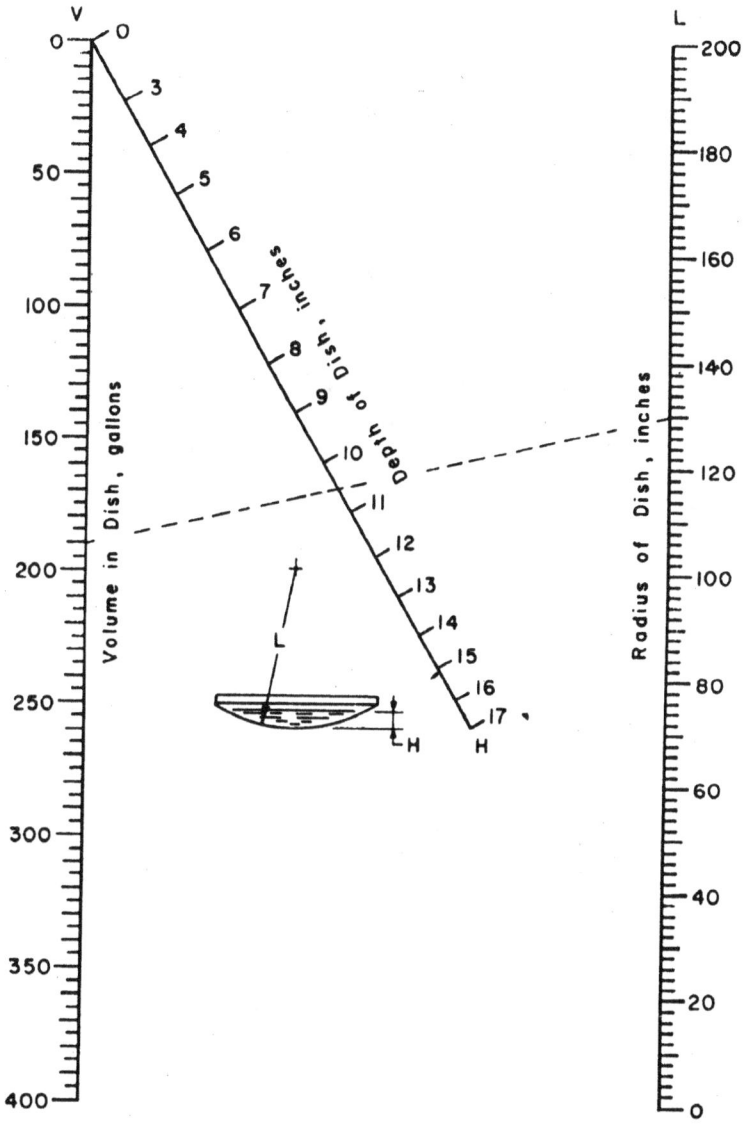

Figure 8-40

8-41   Poly-Plant Conversion

R. L. BARRETT

When converting propylene and butylene to poly gasoline in a catalytic polymerization unit in an oil refinery it is necessary to calculate the degree of conversion of the olefin feed.

The accompanying nomograph, Figure 8-41, is useful for making this calculation rapidly. It is based on the following equation:

$$C = \left[\frac{1 - \dfrac{SG}{FF}}{100 - SG} \right] 10{,}000$$

where   C   = % conversion
         FF = % olefins in fresh feed
         SG = % olefins in spent gas from reactors

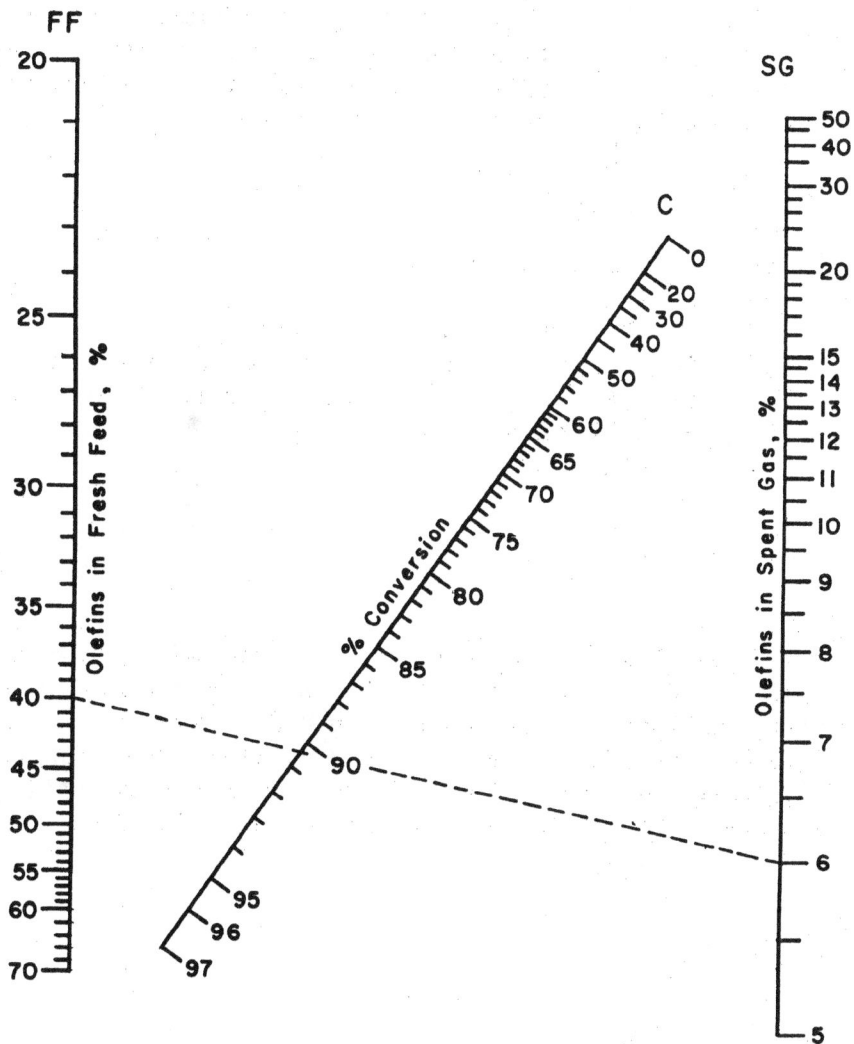

Figure 8-41

8-42   Grashof Number

GEORGE M. MACHWART

The magnitude of heat-transfer effects is measured by dimension-less groups. These are also used as a criterion of dimensional analysis, especially in connection with model and scale-up studies. The dimen-sionless group used in natural convection is the Grashof Number:

$$N = \frac{D^3\, d^2\, \beta g\, \varDelta t}{\mu^2}$$

where D = pipe diameter, ft
       d = density, lb/ft^3
       g = acceleration of gravity, ft/(hr)^2
      β = volumetric coefficient of expansion, (ft^3/ft^3)/°F
     $\varDelta t$ = temperature difference between fluid and wall, °F
     μ = Viscosity, lb/(ft)(hr)

Typical Example

To use Figure 8-42, draw a straight line from 100 on the d-scale to 1000 on the β-scale to intersect R_1. From the intersection on R_1, draw a line to 10 on the $\varDelta t$-scale to cross R_2. From this intersection, draw a line to 0.1 on the D-scale to intersect R_3. Finally, from this R_3 intersection, draw a line to 100 on the μ-scale to cross the N-scale at 4.3×10^9. Use of the formula results in a value of 4.2×10^9.

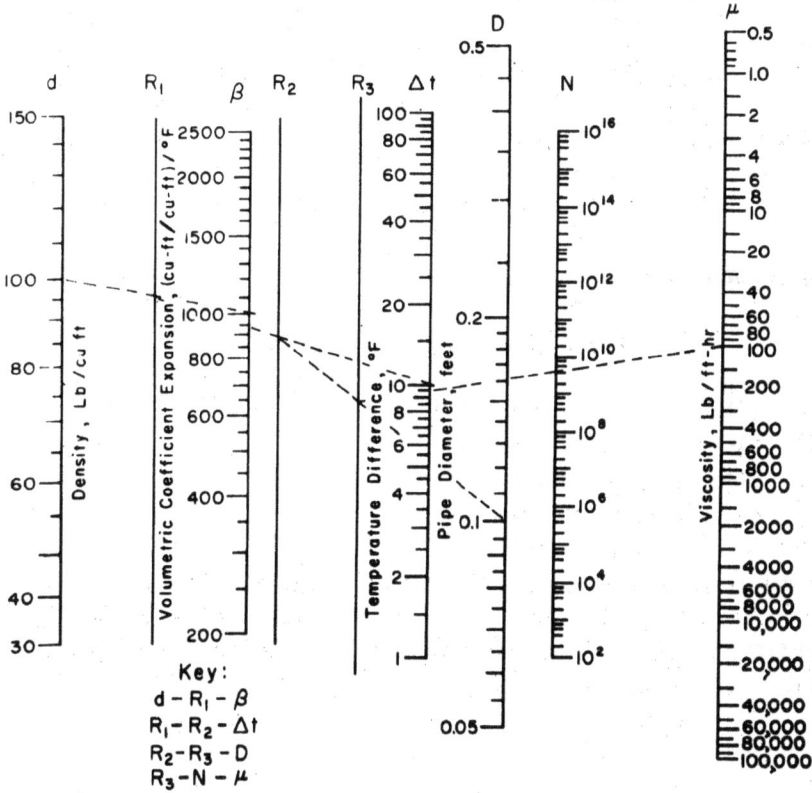

Figure 8-42

8-43a   Characteristics of Formed
Pressure-Vessel Heads

R. D. BIGGS

Figure 8-43a gives a convenient method of estimating characteristics of three commonly used types of formed heads for pressure vessels according to the manufacturer's formulas:

Standard flanged, dished head

$$IDD = \frac{OD}{7}, \quad A = 0.87\,(OD)^2, \quad V = 0.43\,(OD)^3$$

ASME flanged, dished head

$$IDD = \frac{OD}{6}, \quad A = 0.92\,(OD)^2, \quad V = 0.59\,(OD)^3$$

Ellipsoidal head

$$IDD = \frac{ID}{4}, \quad A = 1.19\,(ID)^2, \quad V = (ID)^3$$

where IDD = inside depth of dish, ft
      OD = outside diam, ft
      ID   = inside diam, ft
      A   = surface area, ft²
      V   = volume, ft³

These formulas do not include a straight flange. This can be calculated as part of cylindrical shell. In the absence of tables of head characteristics, the engineer can estimate properties to within approximately 5% with this nomograph.

Typical Example

To find the head characteristics of a 2 ft ID ellipsoidal head, connect the inside diameter with each of the gage points for the ellipsoidal heads (labeled A, V and IDD) in turn and extend to the respective scales to find the area, volume, and inside depth of the dish.

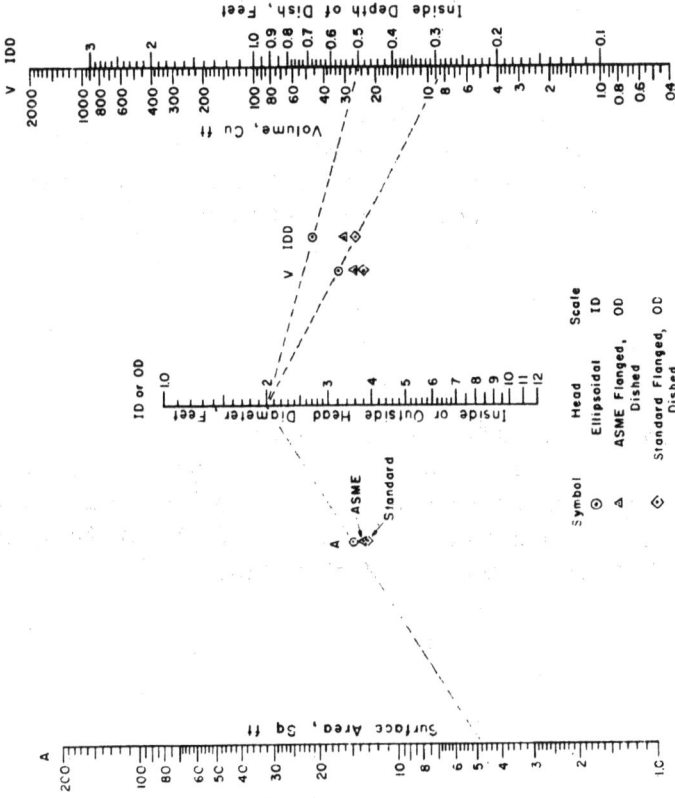

Figure 8-43a

8-43b   Specific Humidity

W. L. PASSMORE

The purpose of Figure 8-43b is to find the molecular weight or the specific humidity of a mixture of air and water vapor when the temperature, pressure, and relative humidity of the mixture are known.

The following formulas were used to construct the nomograph:

$$MW = 28.96 \left[\frac{P_m - P_v}{P_m} \right] + 18 \left[\frac{P_v}{P_m} \right]$$

$$SH = \frac{18 P_v}{28.96 (P_m - P_v)}$$

where P_m = pressure of mixture, lb/in.^{2} abs.

P_v = partial pressure of water vapor, lb/in.^{2} abs.

MW = molecular weight of a mixture of air and water

SH = specific humidity, lb of water per lb of dry air.

Typical Example

For saturated air at 100°F and 14.7 lb/in.^{2} abs., draw a line from 100°F through 100% relative humidity to the reference line. A line from this point on the reference line through 14.7 lb/in.^{2} abs. shows that the molecular weight of the mixture is 28.30 and the specific humidity is 0.043 lb of water per lb of dry air.

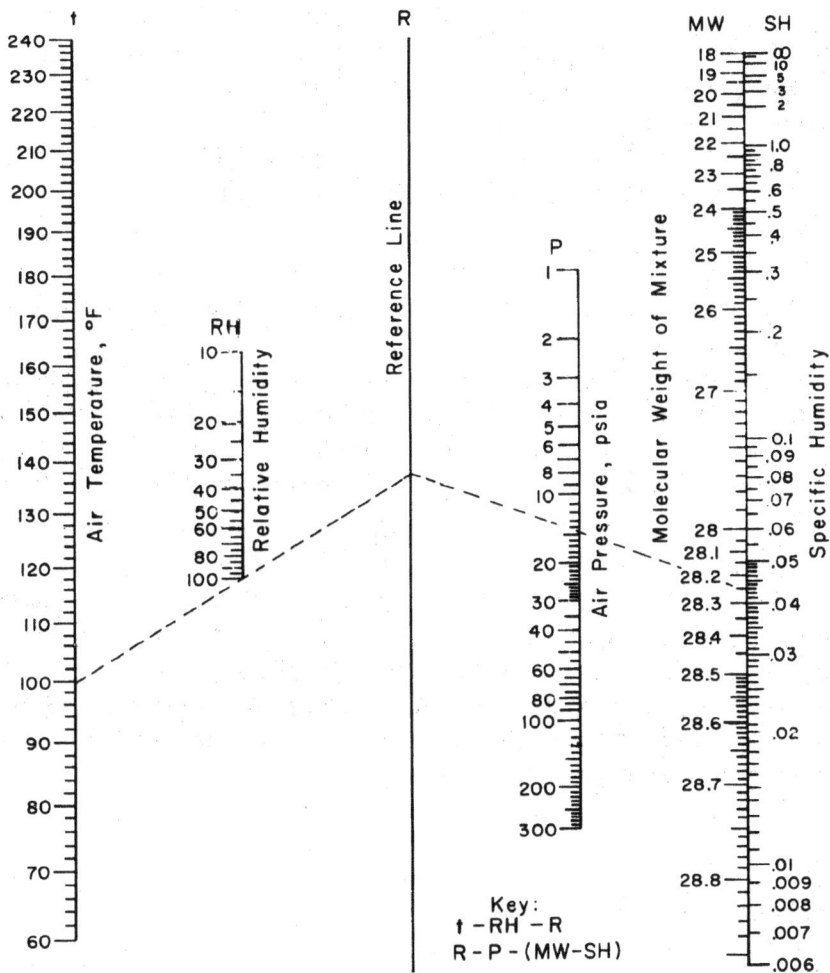

Figure 8-43b

8-44   Hydroxyl and Acetyl Values

GEORGE E. MAPSTONE

Hydroxyl value is used as a measure of the hydroxyl content of an oil, fat or wax. It is defined as the number of milligrams of potassium hydroxide equivalent to the hydroxyl content of the sample and is based on the original sample. The value is calculated from the increase in the saponification number (A.O.C.S. Official Method Cd) after acetylation.

If the saponification value is S before acetylation and S' after acetylation, the hydroxyl value H can be shown to be

$$H = \frac{S'-S}{1.000-0.00075\,S'}$$

The correction factor in the denominator allows for the increase in molecular weight on acetylation.

Acetyl value is defined as the number of milligrams of potassium hydroxide required to neutralize the acetic acid obtained by saponifying one gram of an acetylated oil, fat or wax. (A.O.C.S. Official Method Cd 4-440). It is related to the hydroxyl value by the formula

$$A = \frac{H}{1.000-0.00075\,H}$$

Figure 8-44 has been designed to calculate the hydroxyl and acetyl values from the saponification values before and after acetylation. It can also be used to estimate one saponification value if the hydroxyl or acetyl number and the other saponification value are known. Such a relationship is often required to allow the use of correct quantities of reagents for analysis.

Typical Example

A sample is expected to have a hydroxyl value of about 220 and a saponification value of about zero. What is the expected saponification number after acetylation? Use a straight edge to connect 220 on the H-scale with zero on the S-scale, cutting the S'-scale at 190 (calculated value 189). Thus, the sample should have a saponification value of about 190 after acetylation. Also, opposite 220 on the H-scale, read the acetyl value on the A-scale as 189.

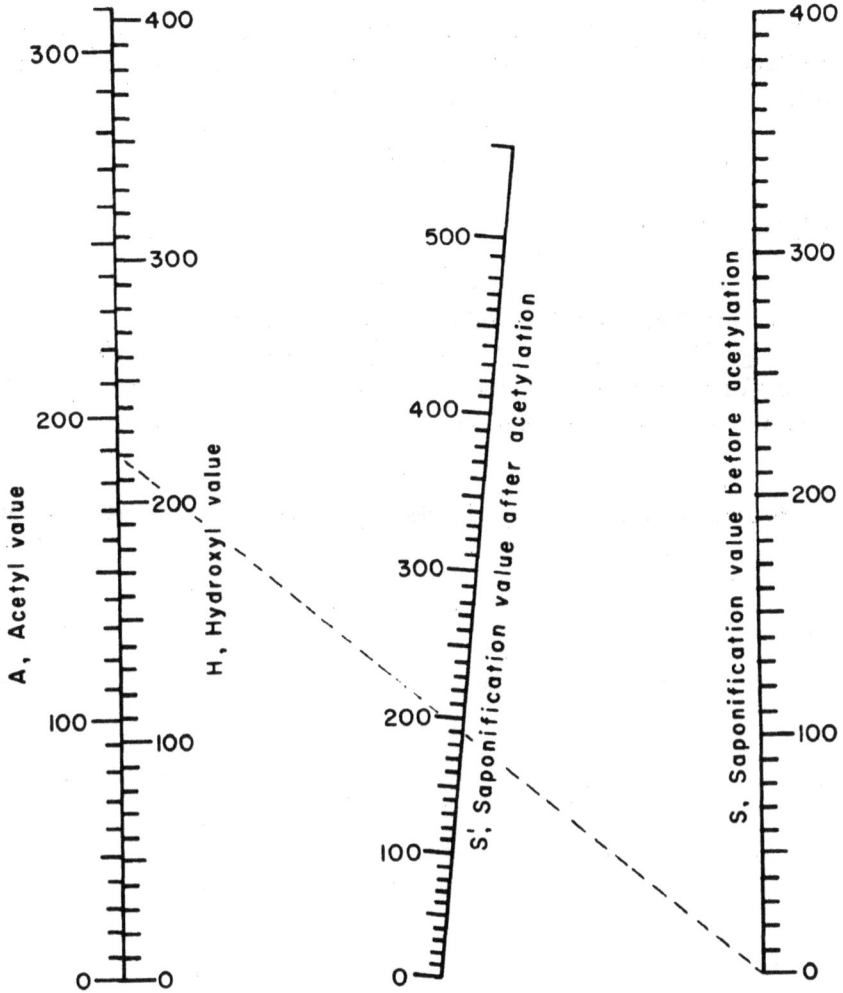

Figure 8-44

8-45   Solubility of Fatty Acids in Benzene

GEORGE E. MAPSTONE

Published data[1] for the solubility of normal fatty acids in anhydrous benzene are presented in convenient nomographic form in Figure 8-45.

Typical Example

How much palmitic acid will be dissolved in benzene at 38°C? Draw a straight line from 38 on the temperature scale through the circled point for palmitic acid, and read the solubility as 45% by weight of acid on the concentration scale.

[1]Markley, K. S., *Fatty Acids*, Interscience Publishers, New York, 1947, p. 191.

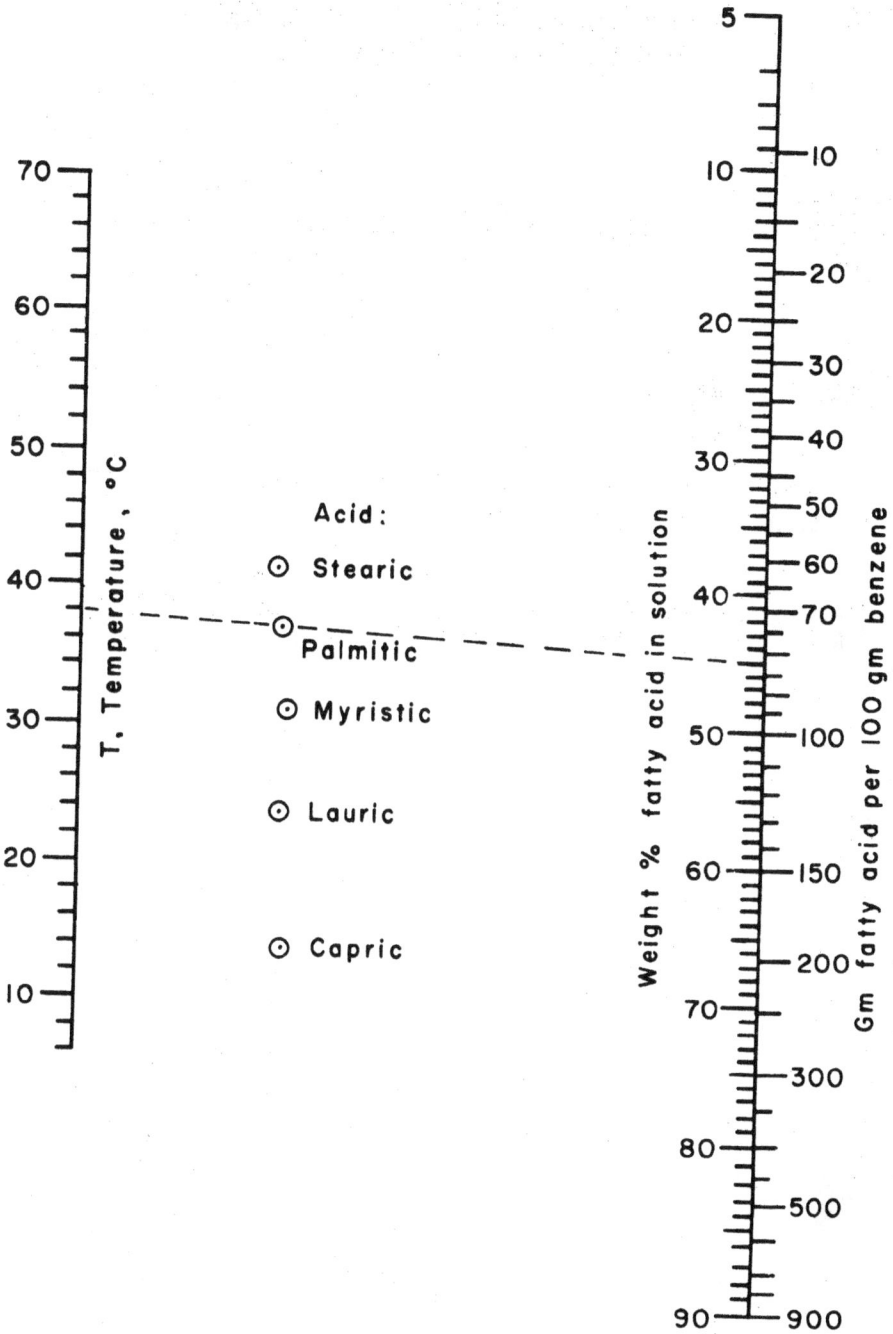

Figure 8-45

8-46   Solubilities of Higher Fatty Acids
in Methanol and Ethanol

GEORGE E. MAPSTONE

Figure 8-46 presents published solubility data[1] for lauric, myristic, palmitic, and stearic acids in methanol and 95% ethanol as a function of temperature. Data are presented in the chart with an accuracy of 1°C.

Typical Example

The dotted line shows that the solubility of stearic acid in 95% ethanol at 54°C is 60% by weight of solution.

[1]Ralston, A. W., and Hoerr, C. W., *J. Org. Chem.*, **7**, 546-555, 1952; Markley, K. S., *Fatty Acids*, 193-195, Interscience Publishers, New York, 1947.

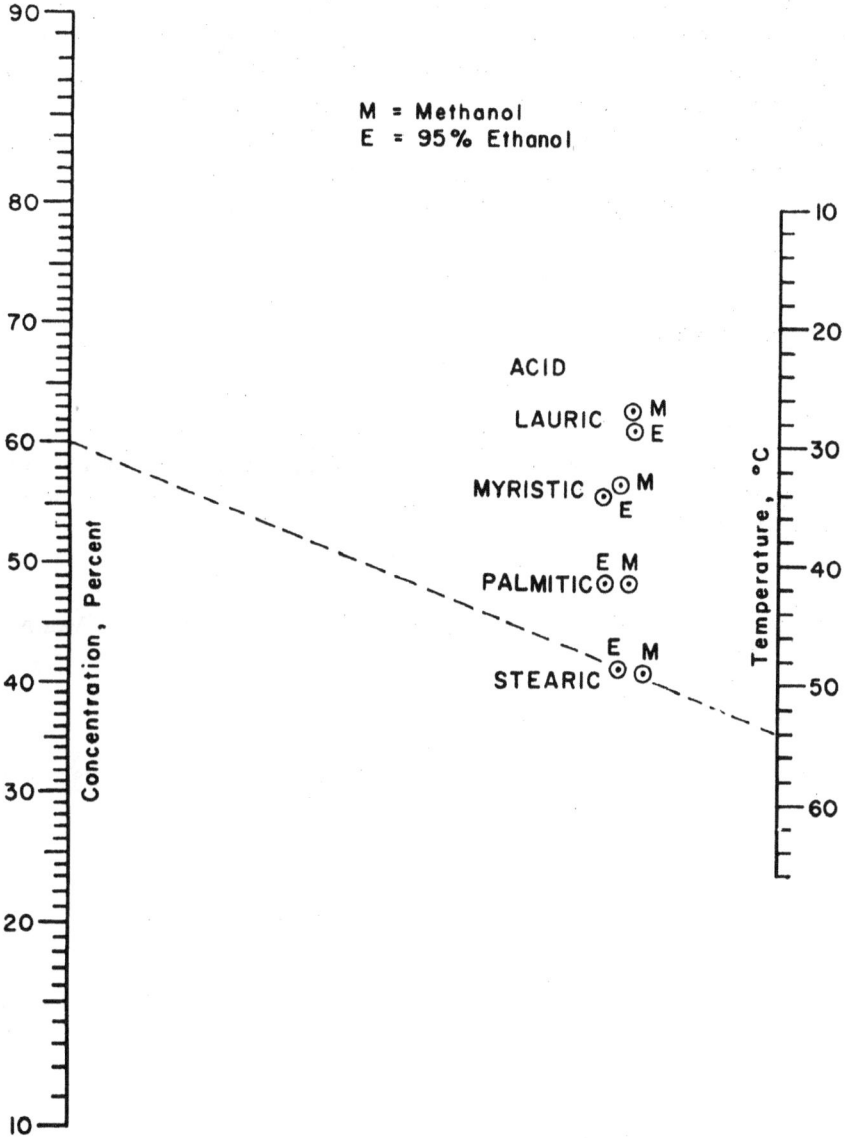

Figure 8-46

8-47   Solubility of Hydrogen in Water

GEORGE E. MAPSTONE

Published data[1] for the solubility of hydrogen in water at high pressure are presented in Figure 8-47. The solution of hydrogen in water obeys Henry's law. The original data on which this chart is based agree with other published low-pressure solubilities when extrapolated to a pressure of 1 atm. This chart can therefore be used safely down to pressures of 1 atm. by multiplying the pressure and solubility scales by some suitable power of ten.

Typical Example

The dashed line shows that, at 20°C and 80 atm pressure, 1.43 cm³ of hydrogen (measured at 0°C and 760 mm) dissolves in 1 g of water. Similarly, at 8 atm the solubility is 0.143 g/cm³, while at 0.8 atm it is 0.0143 g/cm³.

---

[1] A. Seidell, *Solubilities of Inorganic and Metal Organic Compounds*, Van Nostrand, New York, 3rd ed., Vol. 1, p. 554, 1940.

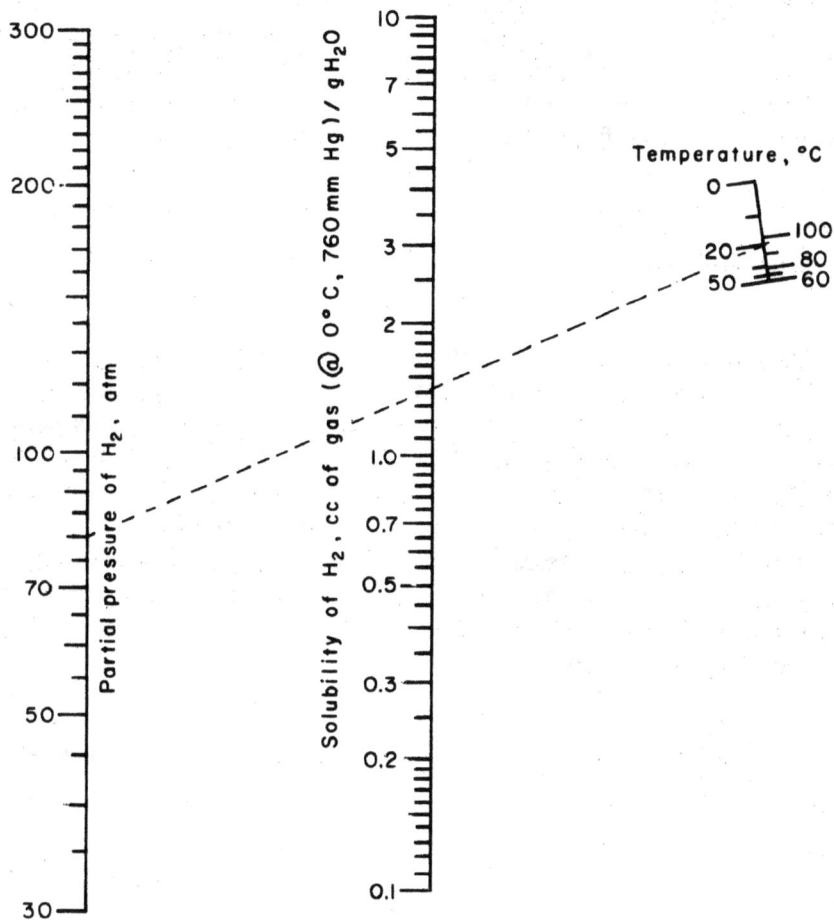

Figure 8-47

8-48   Mutual Solubilities of Light Hydrocarbon-Water Systems

GEORGE E. MAPSTONE

Data on mutual solubilities of light hydrocarbon-water systems are useful in studying behavior of petroleum products in contact with water. These data have been reported by a number of investigators.[1]

Figure 8-48 covers methane, ethane, propane, and n-butane, with water at pressures from 1000 to 10,000 lb/in.[2] abs.

Typical Example

What is the solubility of methane in water at 100°F, at a pressure of 2000 lb/in.[2] abs.? From the key (see nomograph) note that point 1 is the reference for methane in water at 100°F. Connect this point with 2000 on the pressure-scale and read the mole fraction as 0.0019 on the right-hand scale. Under these conditions, methane will dissolve in water until a mole fraction of 0.0019 is reached.

Note: This chart yields values within one per cent of smoothed original data for propane and butane. Maximum discrepancies with methane and ethane are 5%. These differences appear to be within the accuracy of the original data.

[1]Brooks, Gibbs, and McKetta, *Petroleum Refiner*, **30**, (10) pp. 118-120, 1951.

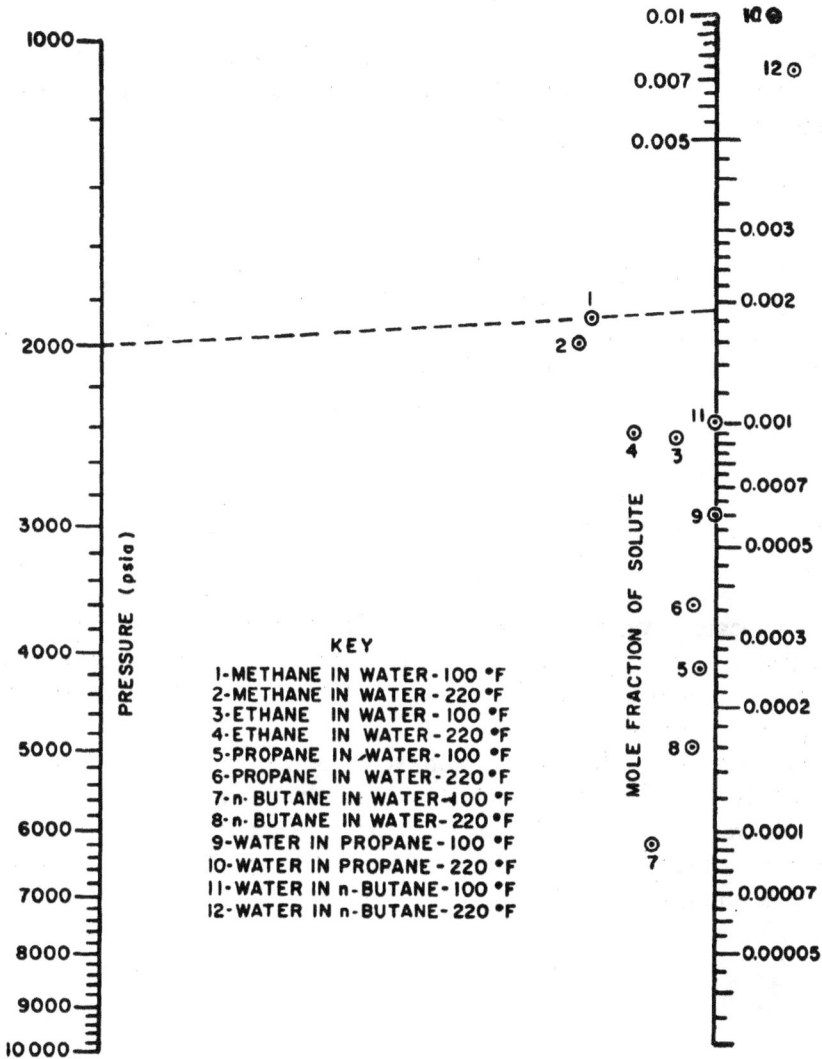

Figure 8-48

8-49   Vapor-Liquid Equilibrium at Constant Relative Volatility

S. M. WALAS

For many binary mixtures the relation between compositions of liquid and vapor phases at equilibrium is expressed on the basis of constant relative volatility by

$$y = \frac{ax}{1-x+ax}$$

where   a   = the relative volatility

x, y   = the compositions of the liquid and vapor phases
in terms of mole fractions. Repeated solutions of this equation are accomplished conveniently with Figure 8-49.

Typical Example

To locate a value of y corresponding to a given value of x at a particular relative volatility, construct a line through 1 on the a-scale and the numerical value of x on the x, y-scale and find the intersection with the reference axis.

Pivoting about this intersection, draw a line to the given relative volatility on the a-scale. Intersection with the x,y-scale then gives the desired value of y. When x=0.25 and a=2.4, y=0.44. A complete x-y diagram for distillation calculations may be obtained quickly in this manner.

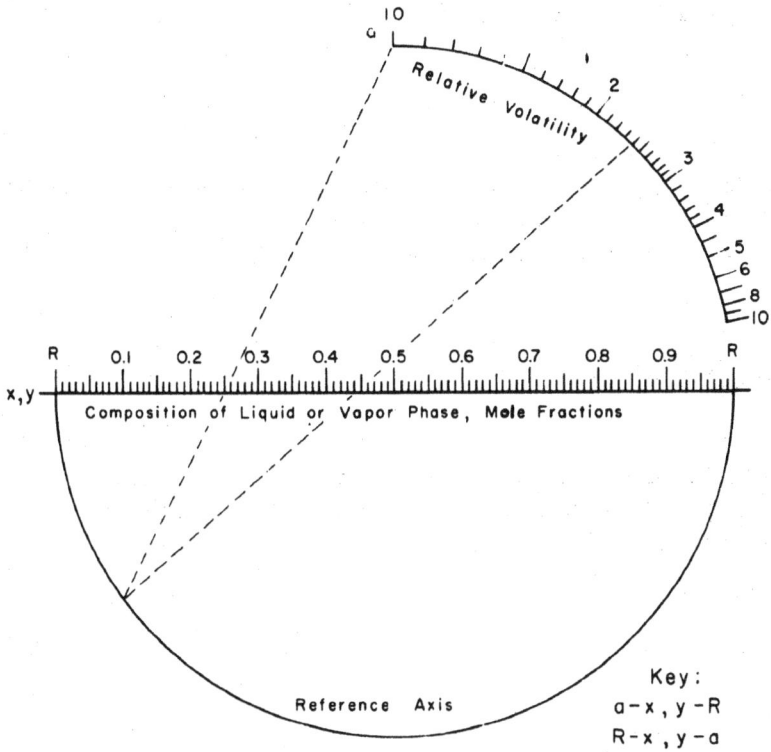

Figure 8-49

8-50   Vapor-Liquid Equilibrium at Constant Relative Volatility

JOSEPH I. LACEY

For use in distillation calculations, Figure 8-50 covers vapor-liquid equilibrium at constant relative volatility, in accordance with the equation

$$y = \frac{\alpha x}{1 - x + \alpha x}$$

where x, y = the mole fractions of the same component in the liquid and vapor phases

α = the relative volatility

Typical Example

On the chart, the broken index line shows that a vapor-phase composition of 0.44 mole fraction is consistent with a liquid-phase composition of 0.25 mole fraction when the relative volatility is 2.4.

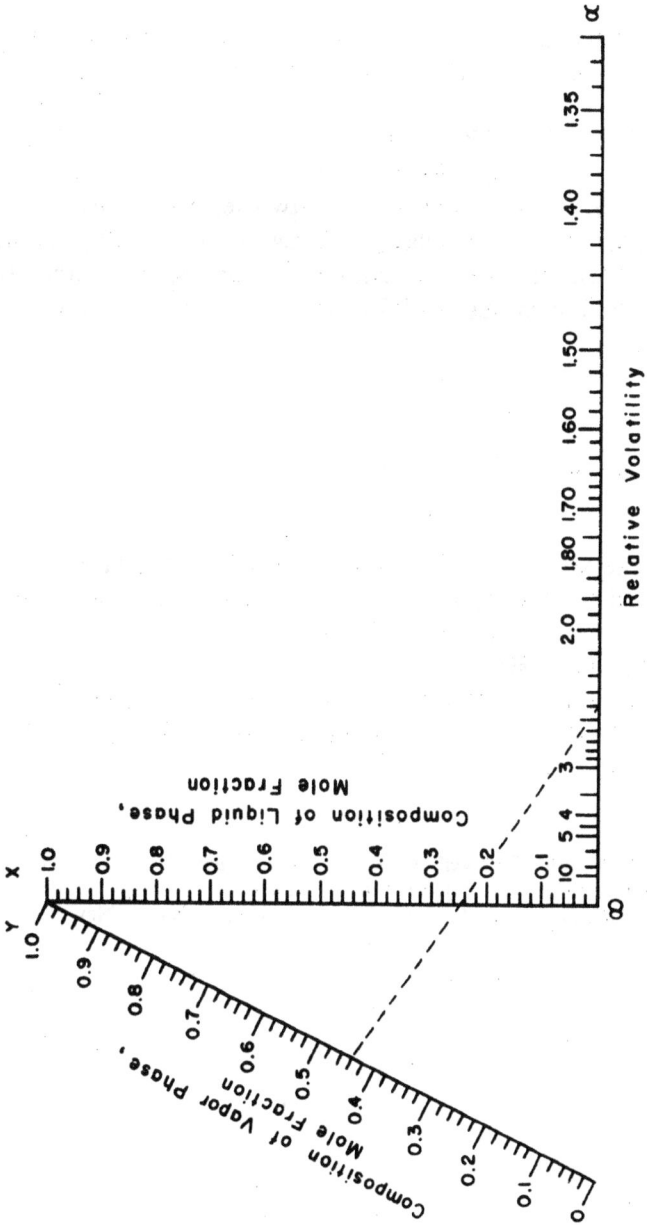

Figure 8-50

8-51   Viscosities of Nitric Acid Solutions

ABE DEVORE

Based on reliable data,[1,2] Figures 8-51a and 8-51b can be used to determine the viscosities of aqueous solutions of nitric acid at various temperatures.

Figure 8-51a is for HNO_3 concentrations of 10-60%. Figure 8-51b covers the 65-100% range. As the viscosity of the acid reaches a maximum value at a concentration of 65%, two separate charts have been constructed for the sake of clarity. Between 60 and 65% strengths, viscosity is constant for all practical purposes, so that either the 60 or 65% calibration can be used for intermediate concentrations within this range.

Typical Examples

What is the viscosity of 30% HNO_3 at 110°F? On Figure 8-51a connect 110°F and 30% HNO_3 with a straight line (not shown) to intersect the viscosity scale at 0.93 centipoise.

What is the viscosity of 80% HNO_3 at a temperature of 122°F? Using Figure 8-51b, draw a straight line from 122 on the T-scale to 80 on the % HNO_3-scale. Extend the line to the Z-scale and read a viscosity of 1.03 cp.

[1]Othmer, D. F. and Cornwell, J. W., *Ind. Eng. Chem.*, **37**, p. 1112, 1945.
[2]Timmermans, Jean, *The Physicochemical Constants of Binary Systems in Concentrated Solutions*, Vol. 4, p. 512, Interscience Publishers, New York, 1960.

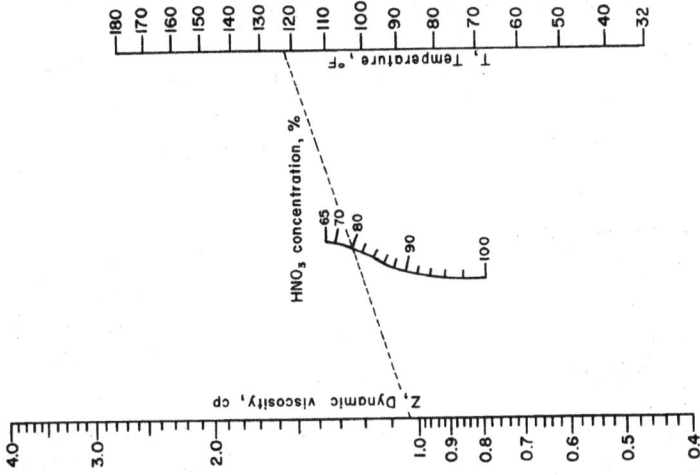

Figure 8-51b

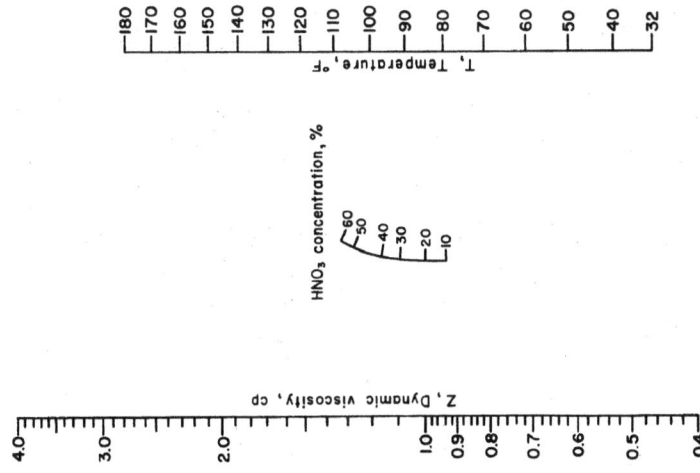

Figure 8-51a

8-52   Gas Viscosities at Low Pressures

R. D. BIGGS

Figure 8-52 presents a convenient method for estimating viscosity of gas at low pressure when no experimental data are available. It is based on an equation developed by Bromley and Wilke:

$$\mu = \frac{0.00333 \, (MT_c)^{\frac{1}{2}} \, F}{V^{\frac{2}{3}}}$$

where M = molecular weight
       T_C = critical temperature, °K
       V = critical volume, cm³/g-mole
       F = tabular function of temperature and molecular compound
       μ = gas viscosity, centipoises

Values of function F can be estimated from a table for given reduced temperature T_R = T/T_C. This equation is limited in use to pressures no greater than a few atmospheres, especially at the lower temperatures. Reed and Sherwood have summarized the best methods for estimating critical properties if these data are not available for a specific compound.

The equation does not hold for hydrogen or helium, and deviations of 12 to 16% are obtained for ammonia and methyl alcohol. However, it predicts viscosities within 5% of experimental values for most gases and vapors.

Typical Example

Estimate the viscosity of ethane gas at atmospheric pressure and a temperature of 50°C. Critical properties and other data are as follows:

$$V_C = 148 \text{ cm}^3/\text{g-mole}$$
$$T_C = 305.3°K$$
$$T = 323.2°K$$
$$T_R = 1.06$$
$$M = 30.1$$

Starting at the T_C scale on the left, connect T_C = 305 with M = 30.1 and extend to reference line A. Connect point on reference line

A with $T_R = 1.06$ and extend to reference line B. Connect this point on reference line B with $V = 148$ and extend to viscosity scale. Read viscosity as 0.0099 centipoise. An experimental value for 50°C is reported to be 0.00998 centipoise.

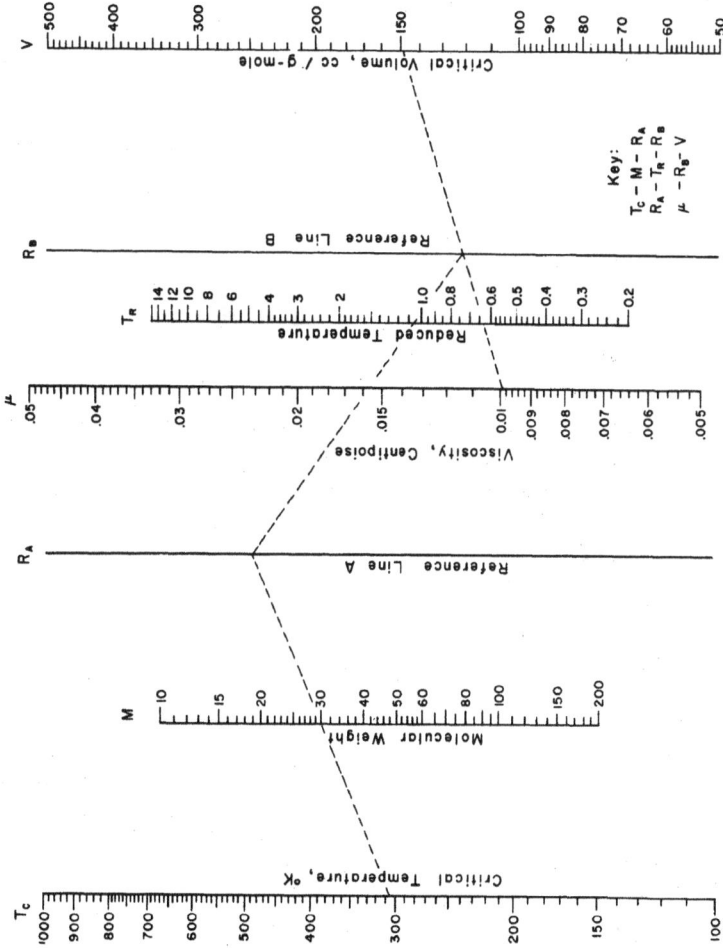

Figure 8-52

8-53   Viscosity of Suspensions

DANIEL S. WILT and HERMAN F. REINHOLD, JR.

In their series of papers on estimating physical properties and thermodynamic data, Johnson and Huang suggested the following equations for estimating the viscosity of a suspension of solids in a liquid for use in fluid-flow and heat-transfer calculations when experimental data are not available:

For volume fraction of solids ≤ 0.1, Einstein's equation

$$\mu = \mu_0 \left(1 + 2.5\, X_D\right)$$

and for $0.5 \leq$ volume fraction of solids 0.9, Hatschek's equation:

$$\mu = \frac{\mu_0}{1 - X_D{}^{0.35}}$$

where X_D = volume fraction of solids in suspension

      μ = viscosity of suspension, centipoises

      μ_0 = viscosity of liquid, centipoises

Although Einstein's equation was derived for dilute suspensions of spherical solids, it has also been applied to emulsions. The accompanying nomographs, Figures 8-53a and 8-53b, provide rapid solutions of the preceding equations. (Any consistent units of viscosity may be used.)

Typical Examples

1) 0.04 volume-fraction of solids is added to a liquid of 10 centipoise viscosity. What will be the resulting viscosity? On the nomograph of Figure 8-53a, connect 0.04 on the X_D-scale with 10 on the μ_0-scale and read 11 cp on the μ-scale.

2) 0.7 volume-fraction of solids is slurried in a liquid of 1-centipoise viscosity. What will be the slurry viscosity? On the nomograph of Figure 8-53b, connect 0.7 on the X_D-scale with 1.0 on the μ_0-scale and read 9 centipoise on the μ-scale.

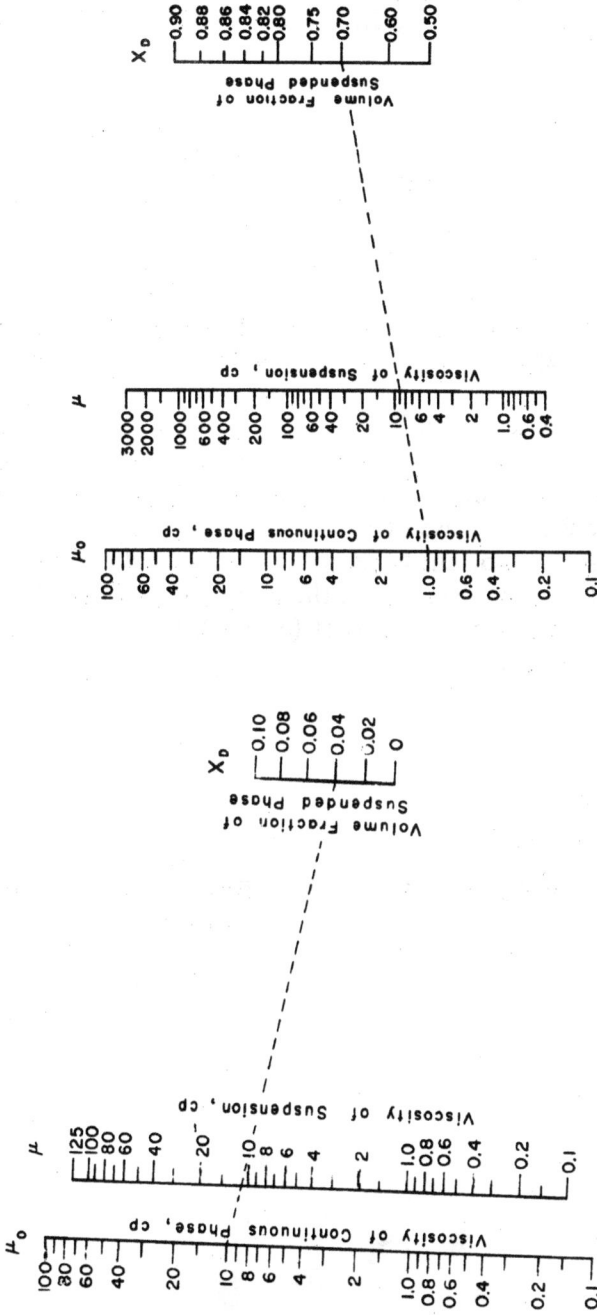

Fig I

Figure 8-53a

Figure 8-53b

8-54   Effect of Entrainment on Fractionation Effciency

ELIZABETH SHROFF

The effect of entrainment on the efficiency of fractionation is represented by the relationship

$$E_a = \frac{E}{1 + \dfrac{e^\circ E}{L/V}}$$

where E_a = efficiency corrected for entrainment, decimal fraction
        E = apparent efficiency of fractionation, decimal fraction
        e° = entrainment, lb/lb of vapor
       L/V = reflux ratio, lb liquid/lb vapor

This equation, which is recommended for use in design calculation by the Research Committee of the AIChE, is presented in nomograph form in Figure 8-54 for ease of handling.

Note: 1. It is important to keep the entrainment ratio well below 20%, to avoid a sharp decrease in the corrected efficiency.

2. When there is no entrainment ($e^\circ = 0$) the entrainment ratio is zero and E_a equals E.

Typical Example

For a reflux ratio of 2.4 and a 0.96 efficiency of fractionation, entrainment of 0.3 lb/lb of vapor results, in a corrected efficiency of 0.857, as shown by the broken line.

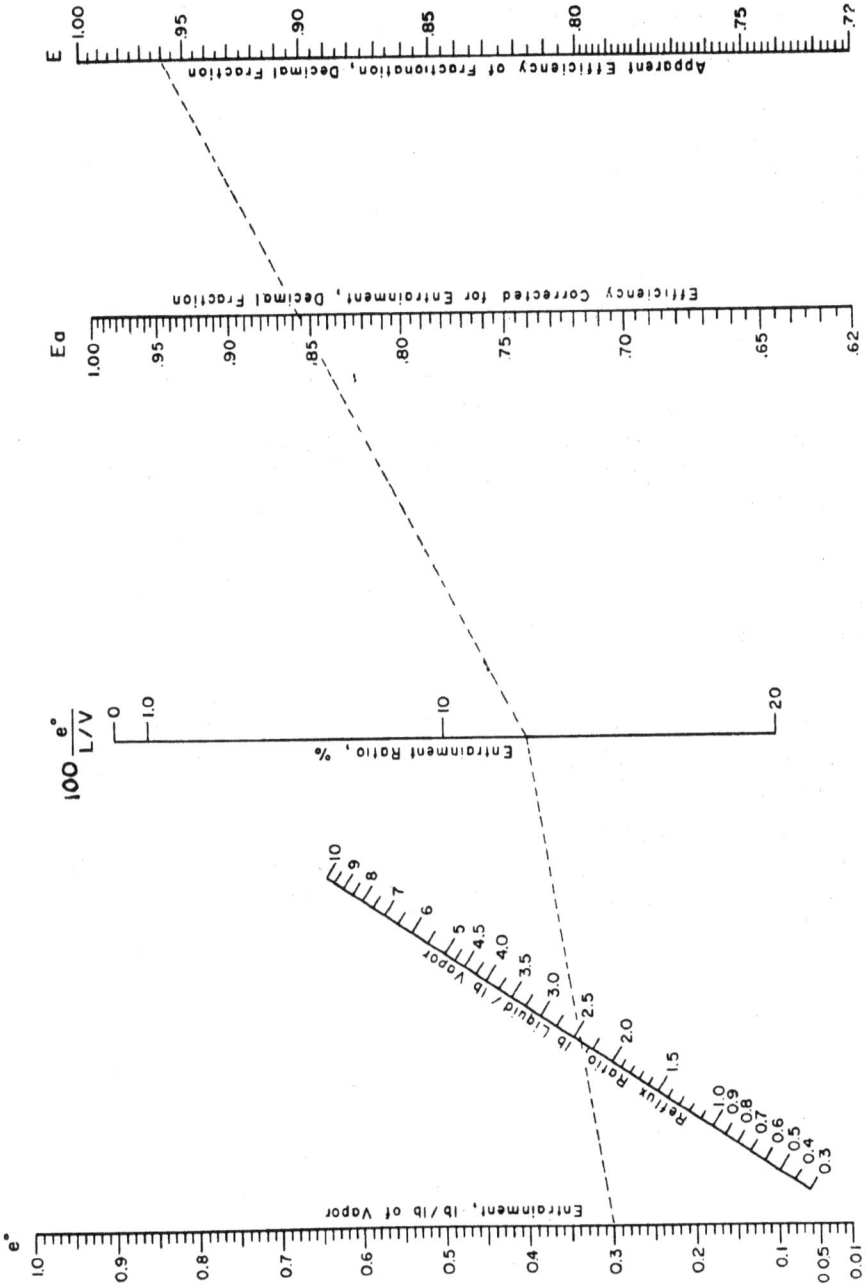

Figure 8-54

8-55   Average Temperature Efficiency of Extended-Surface Heat Exchangers

RONALD A. MUNIER

The average temperature efficiency of an extended-surface heat exchanger is calculated from the equation

$$E_T = I - \frac{A_x}{A_T}(I - E_x)$$

where E_T = average temperature efficiency of the total heat-transfer surface for the side in question

E_x = extended-surface efficiency

A_T = total surface exposed to the side in question, ft^2

A_x = extended surface exposed to the side in question, ft^2

Figure 8-55 provides a rapid solution of the equation.

Typical Example

What is the average temperature efficiency of a heat exchanger having an extended-surface area of 600 ft^2, a total surface area of 1000 ft^2, and an extended-surface efficiency of 0.750?

From 600 on the A_x-scale, draw a straight line to 0.750 on the E_x-scale. Align the intersection this line makes on the R-axis with 1000 on the A_T-scale and read 0.850 on the E_T-scale.

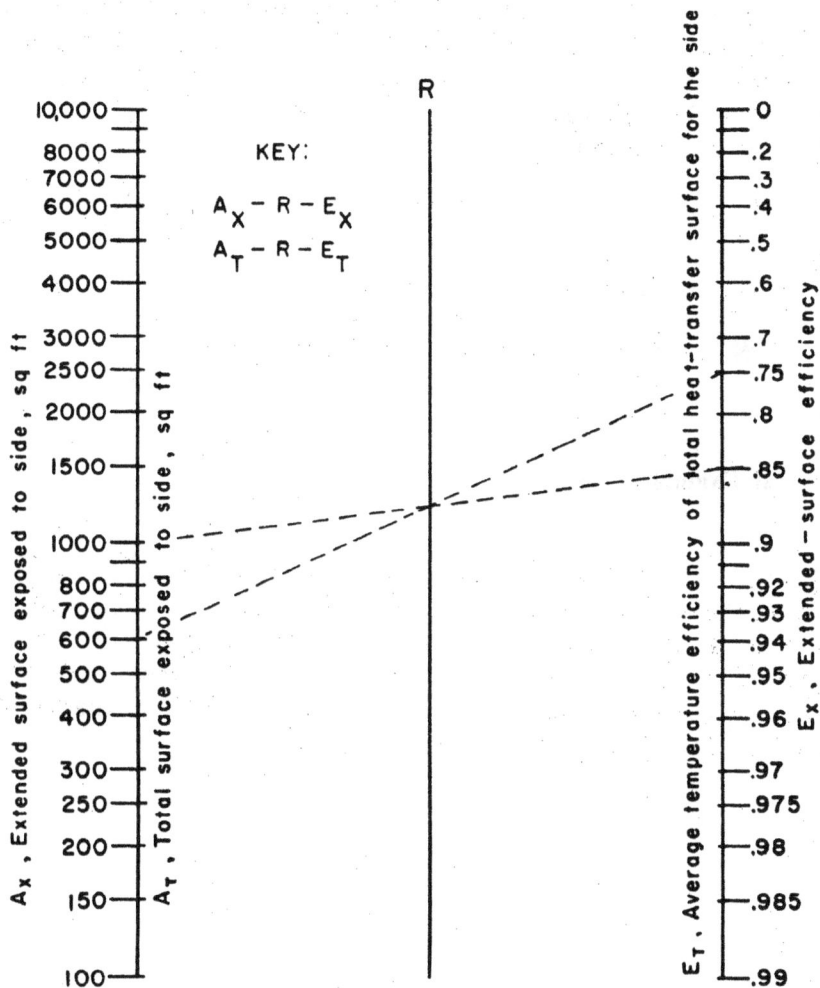

Figure 8-55

8-56   Efficiency of Extraction or Reaction

GEORGE E. MAPSTONE

It is usual to follow and control the efficiency of an extraction process from the analyses of the feed and spent materials. Gas polymerization and alkylation processes, in which one component or group of similar components is reacted and removed from the effluent stream, can be followed in the same manner.

The overall relationship is

$$E = \frac{10000\,(F-S)}{F\,(100-S)}$$

where E = % extraction or conversion
         F $\doteq$ % active material in feed
         S = % active material in discharge

Typical Example

The gas charged to a catalytic poly plant contains 39.0% olefins, and the spent gas contains 19.0% olefins after removal of the polymer. What is the conversion of the olefins in the feed to polymer? Using Figure 8-56, connect 39.0 on the F-scale with 19.0 on the S-scale and extend to cut the E-scale at 63.3%. Therefore, 63.3% of the olefin in the feed gas is converted to polymer.

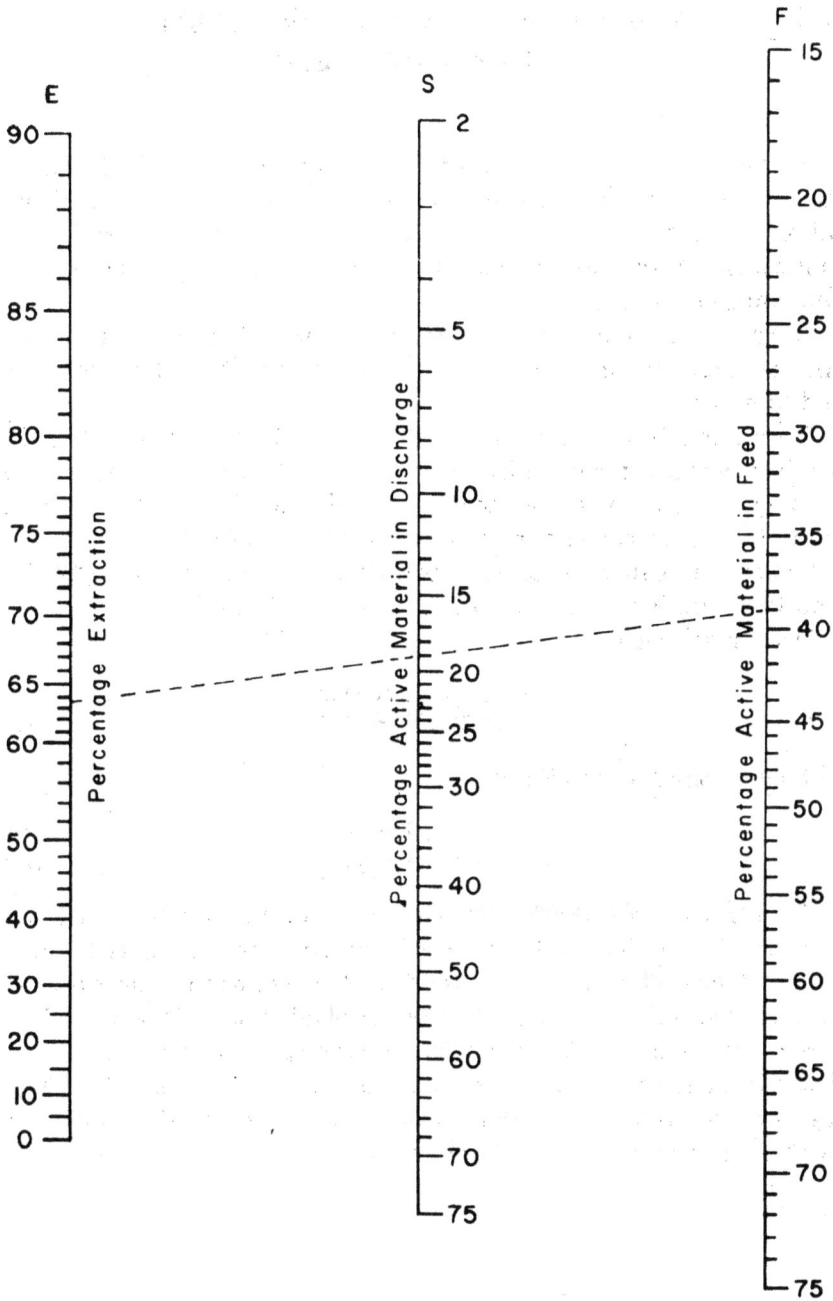

Figure 8-56

8-57   Conversion of Concentration Units for Petroleum Additives

O. M. DUNCAN

A number of ad-ditives are used in petroleum products. These are used to prevent corrosion and carburetor de-icing. They are also utilized as detergents, scavenging agents, gum inhibitors, and valve lubricants. In the heavier fuels, they are used as sludge conditioners, and emulsion breakers.

Several systems are in use to report concentration of additives. Among these are lb/1000 gal of oil, lb/1000 bbl (for 42-gal barrels), and lb/million lb.

For water solutions, ppm are commonly used. For dilute solutions, the density is practically unity, making it easy to convert from volume to weight basis. Variable density of petroleum products makes it more difficult to convert from one system to another.

Figure 8-57 can be used to convert from lb/1000 bbl to ppm for petroleum products ranging from 2 to 82° API gravity.

Two ranges are given:

0 to 250 lb/1000 bbl	(A)
0 to 25 lb/1000 bbl	(B)

with corresponding ranges of

0 to 1000 ppm	(A′)
0 to 100 ppm	(B′)

To calculate the concentration of an additive in ppm when the number of pounds of additive used per 1000 bbl of oil is known, connect the API gravity of oil found on the righthand scale with the number of lb/1000 bbl on one of the left-hand scales. Where the line crosses the oblique scale, read the number of ppm on the appropriate side, depending on which range is being used (A to A′ and B to B′). Similarly, the ppm and gravity scales can be used to determine the number of lb of additive/1000 bbl of oil.

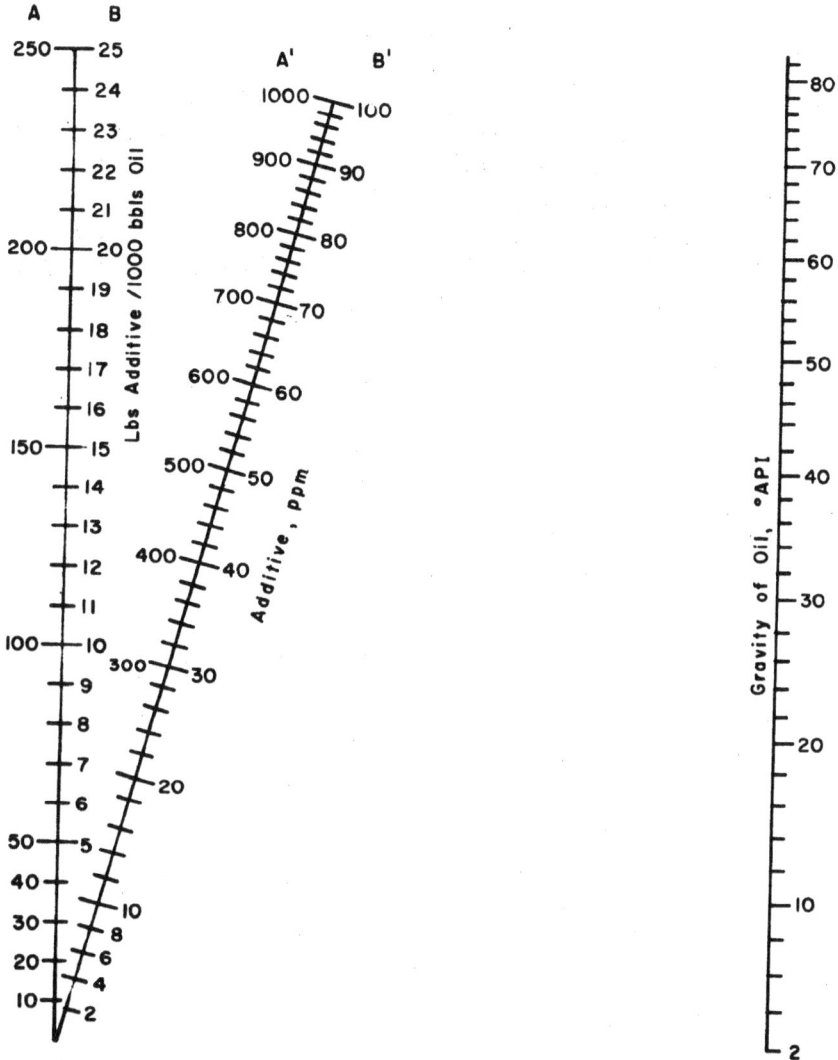

Figure 8-57

8-58   Alkali Equivalents

BRUCE FADER

Figure 8-58 makes possible quick conversion between common alkalis in terms of neutralizing power (chemical equivalents). The outer scale of the chart shows chemical equivalents directly, so that lining up pound equivalents of acid on the outside scale with the center of the circle will produce the number of pounds of each alkali necessary to neutralize the acid. The chart may be used with grams, tons or any consistent units. Values read on the calcium carbonate-scale should be multiplied by 0.99. Where desired values fall off-scale, use decimal multipliers.

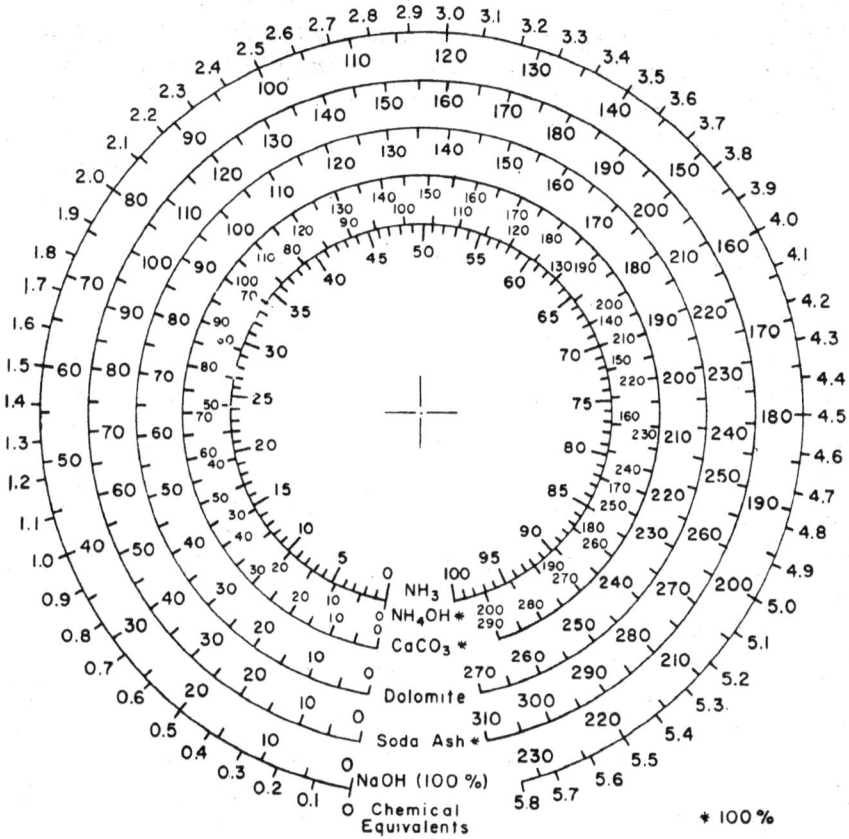

Figure 8-58

8-59   Pressure Conversion

BRUCE FADER

Although conversion of pressure into various units is a simple operation, conversion of a large amount of data is tedious work and prone to error. Figure 8-59 was designed to convert low pressures from one set of common units to any or all of three others.

To use the nomograph simply place a straightedge so that it intersects the known value and lies across the center of the bull's-eye. Readings on all the scales will then be equivalent.

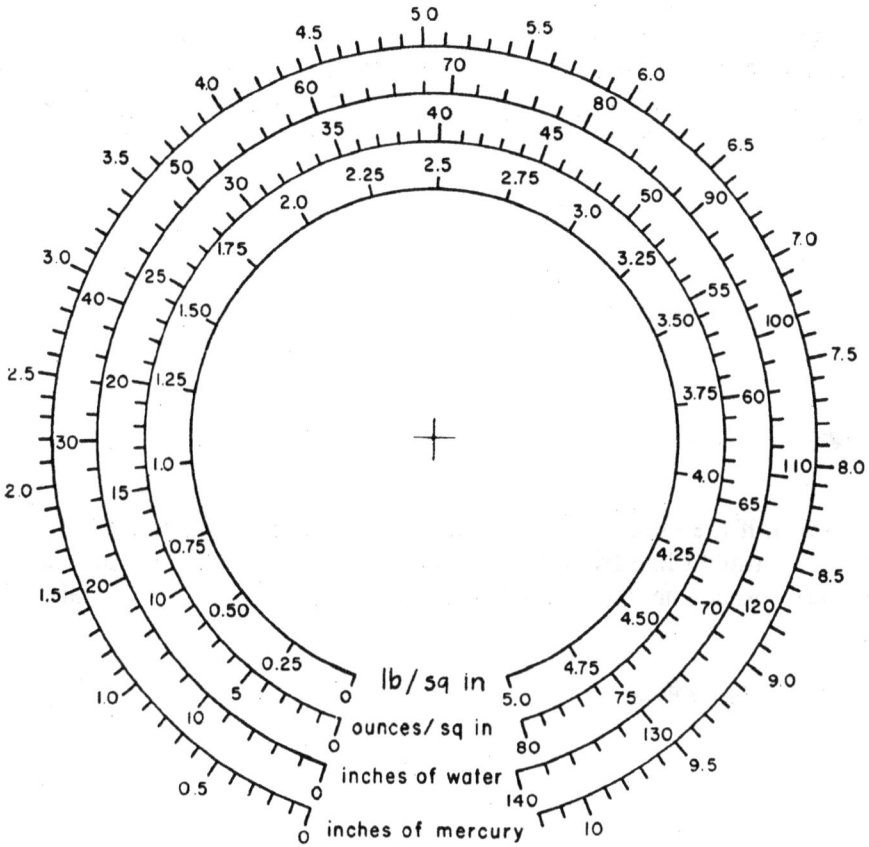

Figure 8-59

8-60   **Specimen-Geometry Factor for Plastics Testing**

W. D. WOLF

A major product test performed on plastics involves the determination of dynamic-modulus data. In obtaining this information, specimen geometry, i.e. the relationship of width to thickness of test samples, is compensated for in various formulas by introducing a factor μ.

Figure 8-60 permits the quick determination of this factor. The chart is based on the equation[1]

$$\mu = 5.33 - 3.36 \frac{b}{a} \left(1 - \frac{b^4}{12a^4}\right)$$

where   a = specimen width
       b = specimen thickness

Typical Example

Find μ if the sample width is 0.250 in. and its thickness is 0.062 in. Draw a straight line from 0.250 in. on the a-scale to 0.062 in. on the b-scale. Read the value of 4.5 where this line intersects the μ-scale.

[1]*ASTM Standards*, p. 417, 1961.

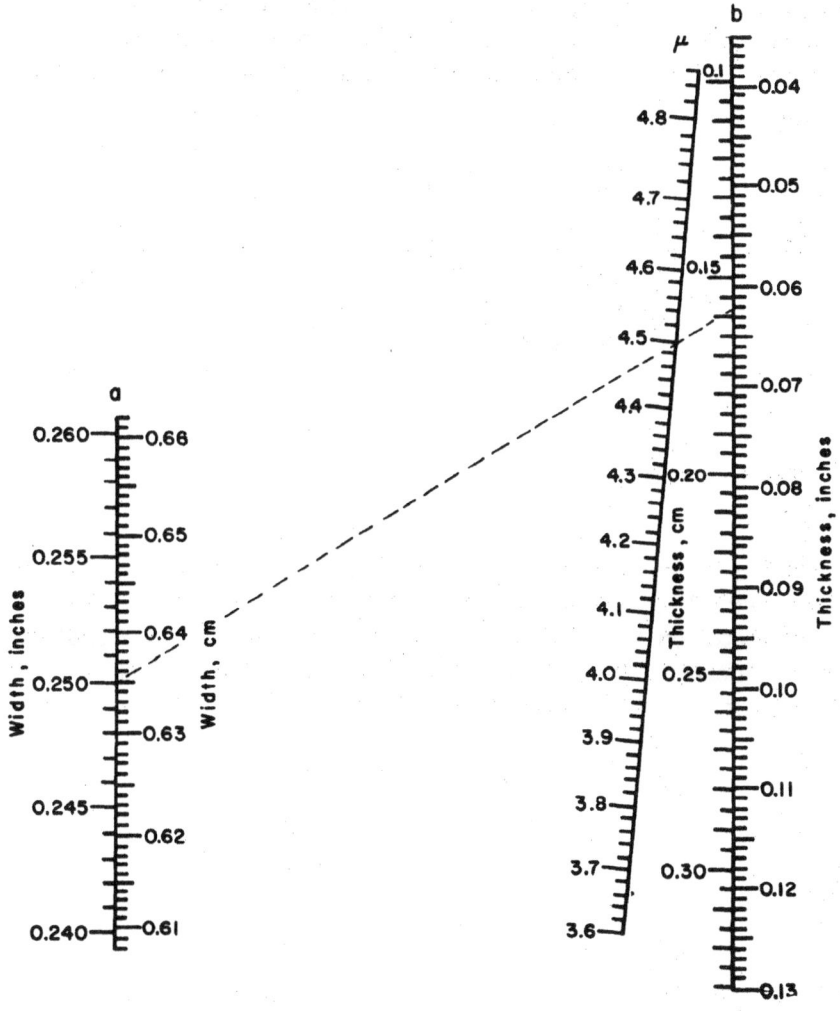

Figure 8-60

8-61   Output from Plastic Extruders

MICHAEL J. GLUCKMAN

Before any type of flat-film extrusion can be commenced, the output in lb/hr must be determined. This quantity depends on speed, web width, and the required film weight in lb/ream (3000 ft^2).

Figure 8-61 permits quick determination of output. It is based on the equation

$$Q = \frac{WBS}{600}$$

where   Q = extruder output, lb/hr
        W = film weight, lb/3000 ft^2
        B = width of film, in.
        S = line speed, ft/min

Typical Example

What is the extruder output required to produce a 40 in.-wide web, coated with 5 lb of film/ream, at a speed of 600 ft/min?

Connect 5 on the W-scale with 40 on the B-scale, and extend the line to intersect the reference scale. Connect this point with 600 on the S-scale, and read the extruder output as 200 lb/hr on the Q-scale.

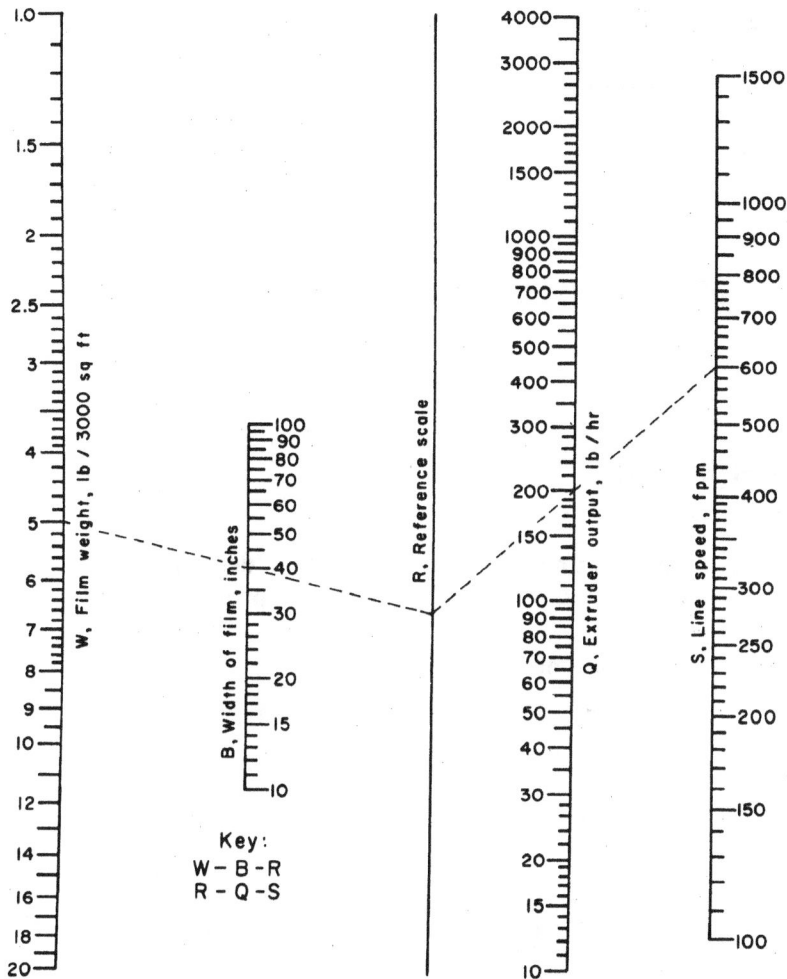

Figure 8-61

8-62   Basis Weight of Plastic Paper Coatings

T. J. GRAIL

The plastic content of a plastic-paper laminate may be expressed in two ways: (a) thickness in mils, and (b) basis weight (lb plastic per ream or per 3000 ft^2).

Thickness is related to the barrier properties desired in the laminate, such as vapor impermeability and water and stain resistance. On the other hand, economy of operation is tied to basis weight — the amount of plastic needed to do the job.

Figure 8-62 relates thickness to basis weight, given a particular density of laminate (in gm/cc). It is based on the equation

$$\text{Basis weight} = 15.65 \times \text{thickness} \times \text{density}$$

Typical Example

For a 0.5-mil coating of cellophane having a density of 1.45 gm/cm^3, the basis weight is found by joining the appropriate points on the outer scales with a straight line. Intersection on the inner scale gives a basis-weight value of 11.5 lb/ream.

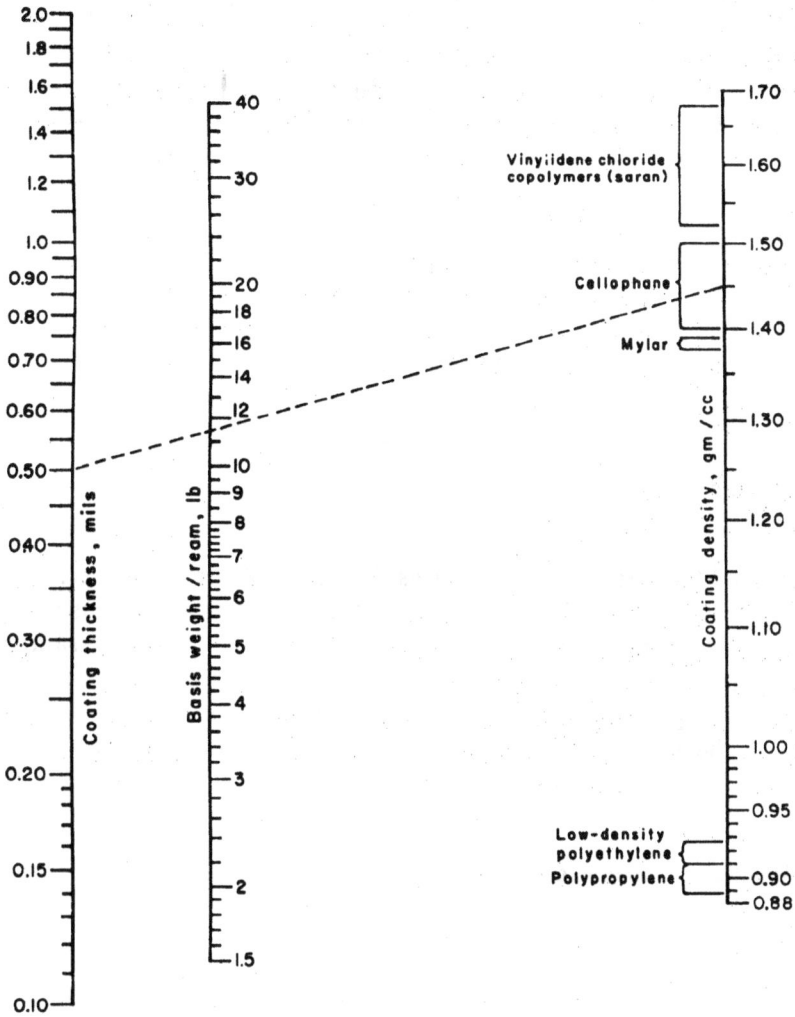

Figure 8-62

8-63   Karl Fischer Determination of Water

G. F. FELLERS and J. P. MESSERLY

The Karl Fischer method[1] for the direct titration of water can be applied to solids, liquids and gases. It is widely used for paints, plastics, petroleum products, carbohydrates, drugs, foodstuffs, explosives, and naval stores. The necessary computation can be readily handled by means of Figure 8-63, which is based on the equation

$$P = 100 \frac{FV}{W}$$

where P = percentage of water
        V = volume of Karl Fisher reagent, ml
        F = concentration, mg water/ml reagent
        W = weight of sample, mg

Typical Example

Use of the chart is illustrated as follows: What is the percentage of water ,when 0.8 ml of Karl Fischer reagent, which contains 6.5 mg of water per ml of reagent, is needed to titrate 800 mg of sample? Following the key, connect 6.5 on the F-scale and 0.8 on the V-scale with a straight line. Note the intersection with the A-axis. Connect 800 on the W-scale with this point, and read the intersection with the P-scale as 0.65% H_2O.

[1]Fischer, K., *Angew. Chem.*, **48**, 394-6, 1937.

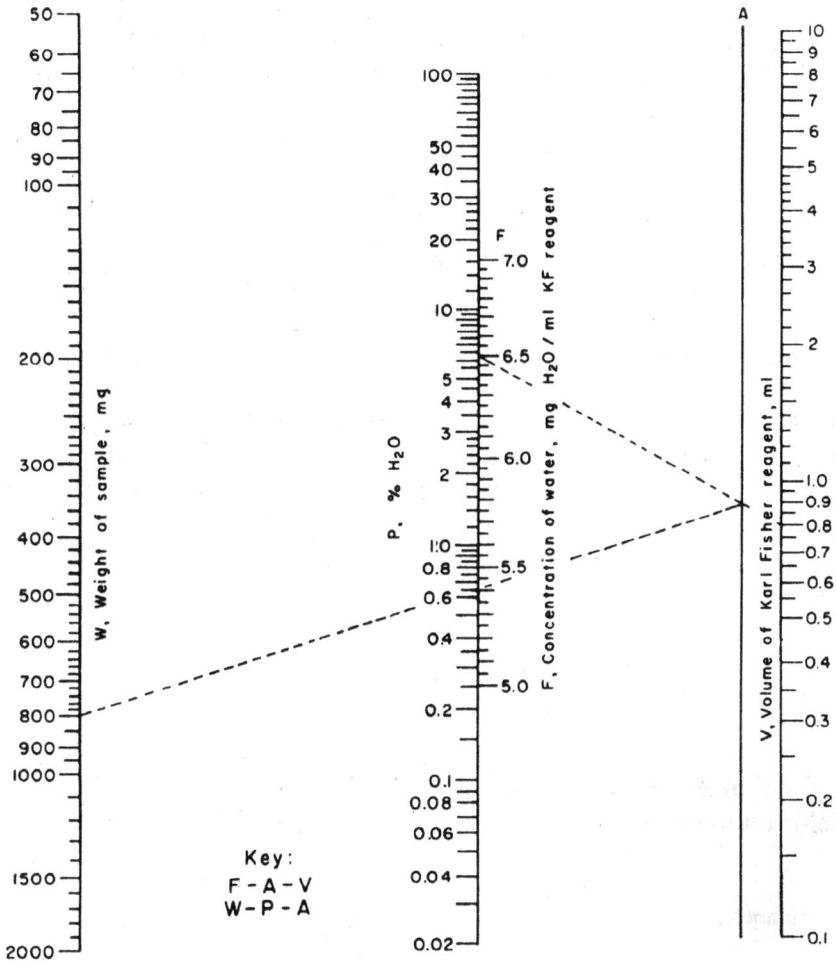

Figure 8-63

8-64   Gas Permeability

THOMAS J. GRAIL

Studies of permeability[1] show that with a given pair of gases the ratio of permeability constants remains roughly the same for a number of dissimilar plastic films — even though the range of actual permeabilities is several-thousand-fold. This leads to the relationship:

$$P = \gamma FG$$

where P = permeability of a given gas/film combination
        F = a constant for the given film
        G = a constant for the given gas
        γ = an interaction coefficient, which is usually close to unity

Figure 8-64 solves this relationship (assuming $\gamma = 1$), by which one can predict a permeability constant for a given gas/film combination, with known F and G values. The nomograph also suggests a possibility of measuring permeabilities with unknown F or G, and in this way establishing the unknown quantity for use in predicting other combinations.

Typical Example

The broken index line shows that the permeability of oxygen through Kel-F fluorocarbon film is 5.0×10^{-10} cm^3/(mm)(cm^2)(sec)(cm Hg).

[1] Stannett and Sware, *J. Poly. Science*, XVI, 81, pp. 89-91, 1955.

Gas permeability

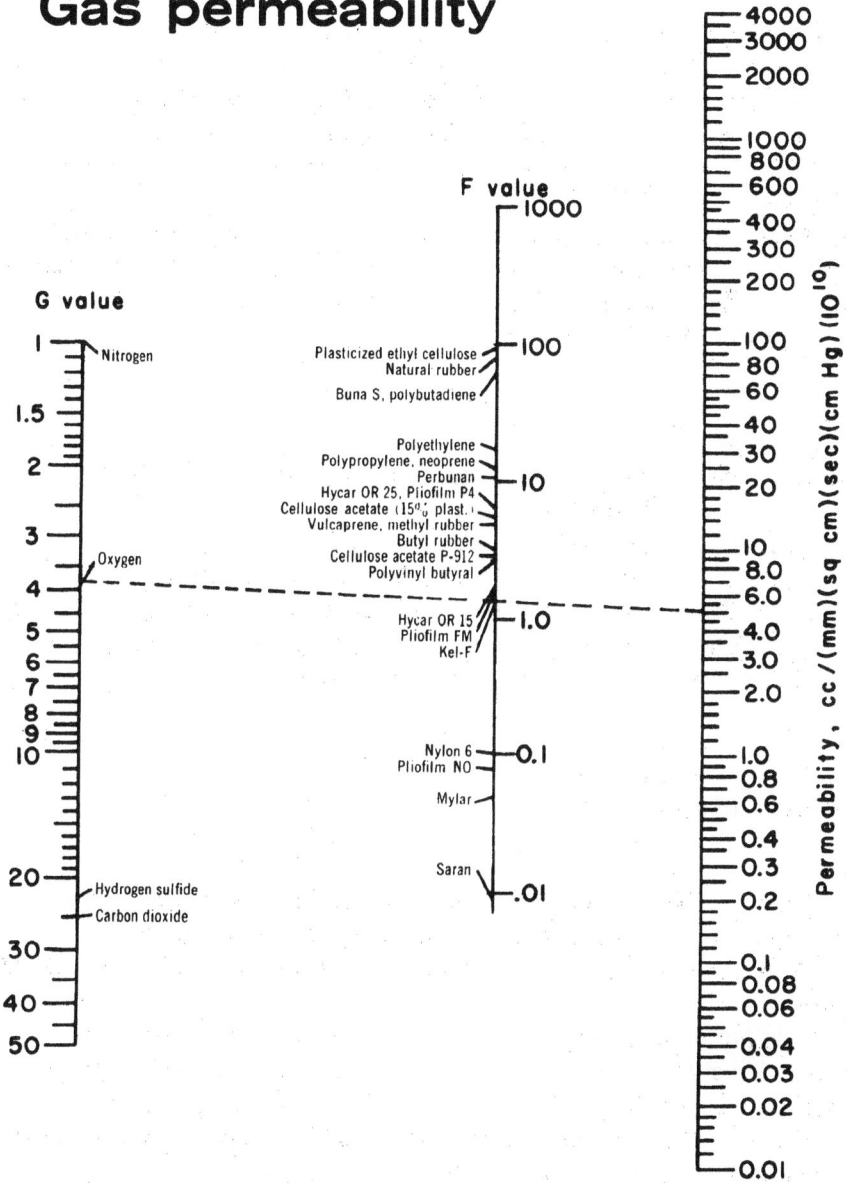

Figure 8-64

8-65   Water Vapor Permeability of Foams

R. R. DIXON

In the measurement of water vapor transmission of plastic foams, ASTM C-355 and E-96 are frequently used.

Basically the procedure is to seal the foam to an impervious container containing a desiccant. This assembly is then stored in a high humidity chamber and is weighed periodically to determine the weight change.

When this change is plotted against exposure time, a curve is obtained which eventually becomes a straight line. Slope of the straight line portion is the rate of flow of moisture through the foam barrier.

The permeability of the foam is expressed by

$$p = \frac{2352\, WL}{At\, \Delta P}$$

where   W = weight of moisture passing through foam, g
        L = foam thickness, in.
        A = area of foam sample, in.^{2}
        t = days during which W was measured
    ΔP = vapor pressure drop across foam, mm Hg
  2352 = conversion constant to give p, perm-inches

This equation is the basis for Figure 8-65.

Typical Example

Foam of 50 in.^{2} is exposed with a 30-mm Hg pressure drop. A straight line connecting these points intersects Index Line 1.

Once the test procedure is established, this point on Line 1 should not vary.

This point connected to a foam thickness of 1.0 inch gives a point on Index Line 2. A line from here to a weight increase of 10 g gives a point on Index Line 3.

This point, connected to a time interval of 3 days, projects to a permeability value of 5.2 perm-inches.

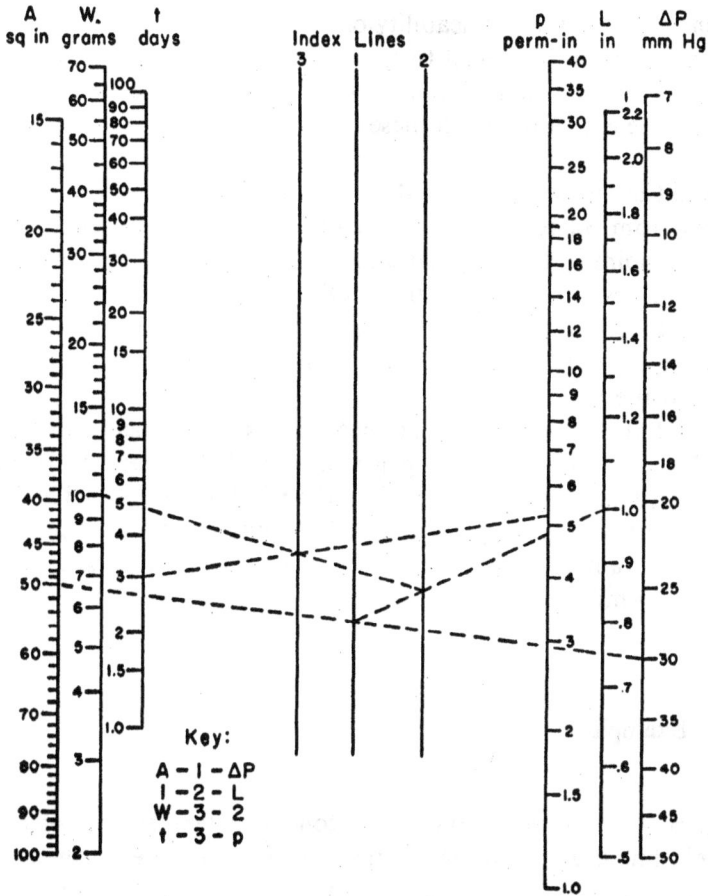

Figure 8-65

8-66   Permeability of Polyethylene to Selected Common Solvents

BILL SCHREMP

Recent work on the permeability of polyethylene to various solvents has produced data of interest to those concerned with the use of polyethylene for packaging and as a vapor barrier. Figure 8-66 serves as a guide to the extent to which these materials will pass through such a film.

It should be noted that these data are the result of tests run at temperatures from 32 to 165°F and in no case were tests made above the boiling point of the material under study. Higher and lower temperature values are extrapolations of the data. Furthermore, the permeability given here is a combined figure for liquid and vapor phases of the solvent. As a rule the liquid will permeate polyethylene faster than the vapor.

It should also be noted that permeability is not a linear function of the thickness of film. Heavier gages of polyethylene tend to be more crystalline (with lower permeability) than lighter gages. There are data[1] showing that in gages of about 100 mils, however, permeability apparently again increases. Data presented here were gathered in tests with 40-mil film.

Typical Example

What is the permeation rate of acetone at a temperature of 60°F? A straight line drawn from 60 on the temperature scale through point 7, for acetone intersects the permeability scale at the desired value, 6 g/(day)(mil)(100 in.²).

[1] J. Pinsky, A. R. Nielsen, and J. H. Parliman, *Modern Packaging*, **27**, (10), 1954.

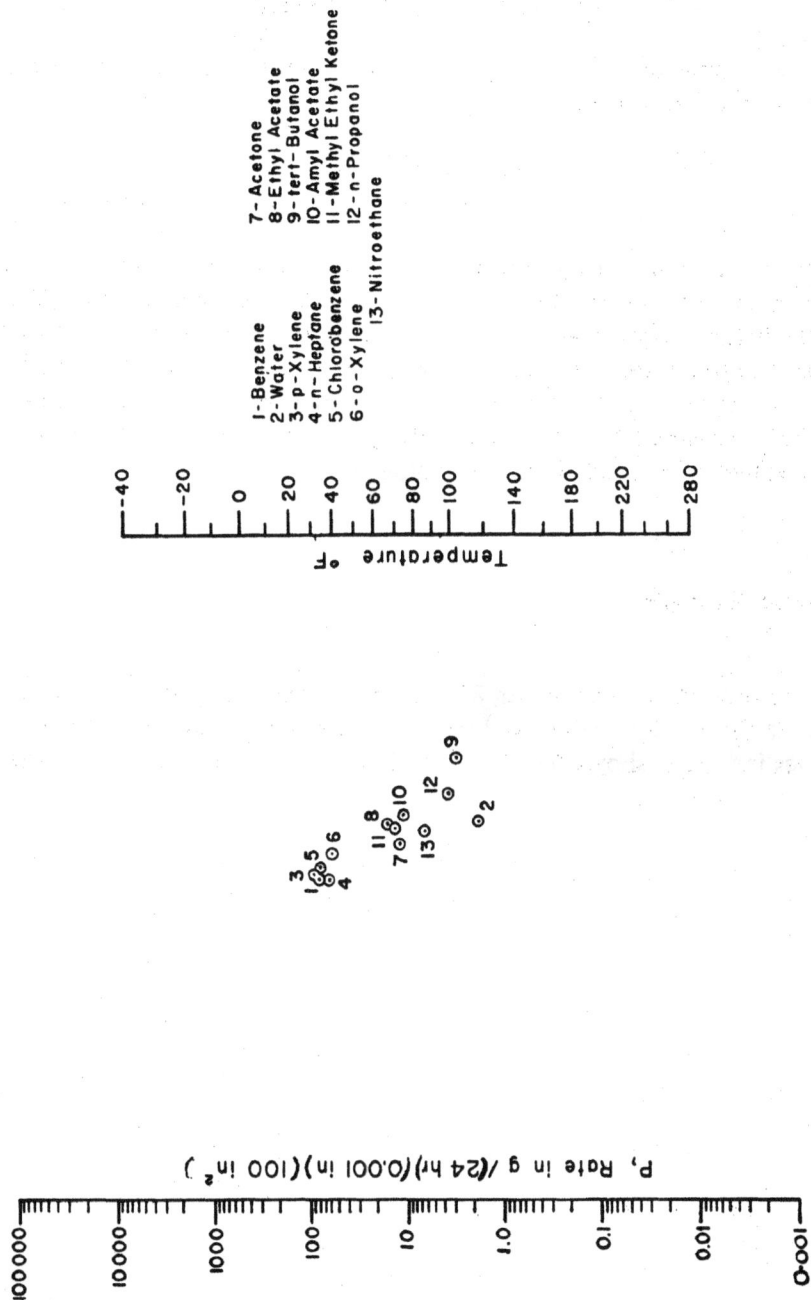

1- Benzene
2- Water
3- p-Xylene
4- n-Heptane
5- Chlorobenzene
6- o-Xylene
13- Nitroethane

7- Acetone
8- Ethyl Acetate
9- tert-Butanol
10- Amyl Acetate
11- Methyl Ethyl Ketone
12- n-Propanol

Temperature °F

P, Rate in g /(24 hr)(0.001 in)(100 in²)

Figure 8-66

8-67   Air Required for a Water Air Lift

CLIFFORD L. DUCKWORTH

The volume of free air required to lift one gal of water has been expressed by the formula

$$V_a = \frac{0.8\, H_t}{C \log \dfrac{H_s + 34}{34}}$$

In this formula, H_t is the total lift in ft (distance from working surface of water to point of discharge), H_s is the running submergence in ft (distance from water level to point of air inlet), and C is a constant, varying with the head required. V_a is the volume of air in ft^3 required to lift 1 gal of water. Variance of C has been incorporated in the interrupted H_t-scale. Accordingly, it is necessary only to connect values of H_t and H_s to determine V_a.

Typical Example

It is desired to raise water 85 ft from a well where the air inlet of the air lift is 75 ft below water level. Connecting the two scales at values indicated shows that 0.58 ft^3 of air will be required to lift 1 gal.

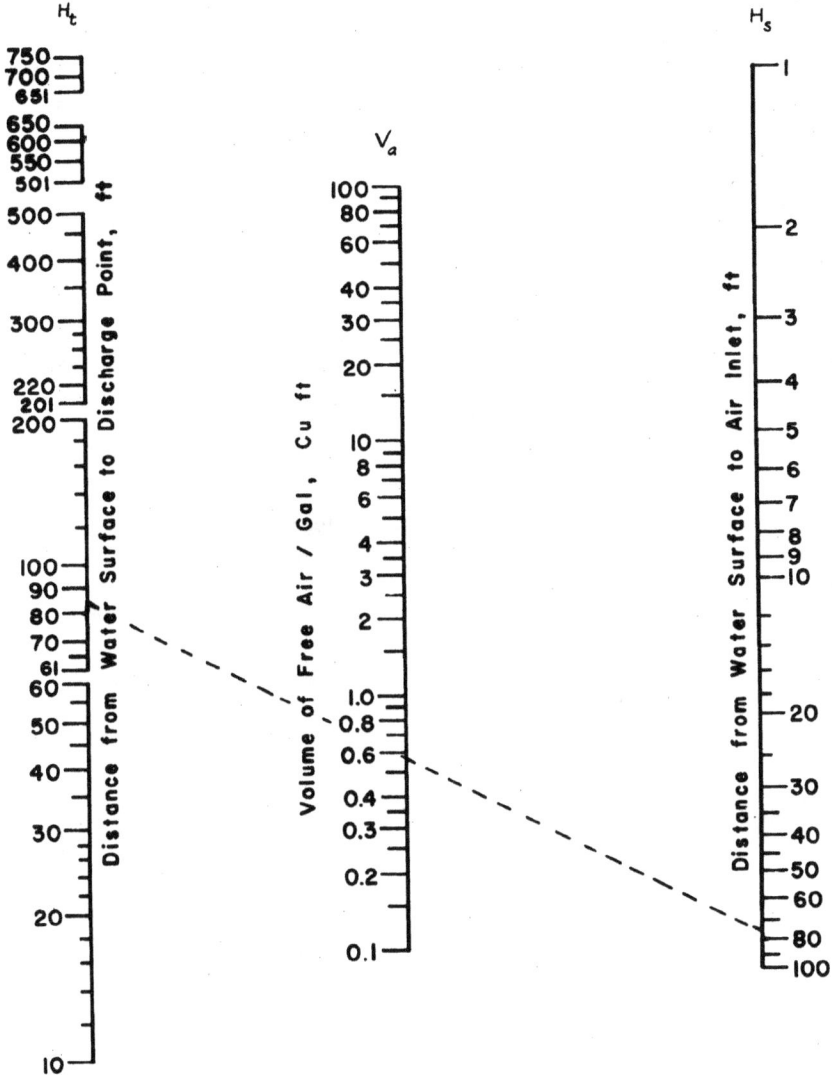

Figure 8-67

8-68   Viscosity of Coal-Tar Distillates

GEORGE E. MAPSTONE

Figure 8-68 provides a quick method for determining the viscosity of any coal-tar fraction at any temperature up to its boiling point.

The basic data[1] shows that coal-tar distillates all have essentially the same viscosity — 0.25 centipoise — at their respective boiling points.

Typical Example

A wash oil of coal-tar origin with an average boiling point of 250°C has a viscosity of 0.72 at 150°C as shown by the broken index line in Figure 8-68.

[1]Briggs, D.H.K., *J. App. Chem.*, 14, 486-489, 1964.

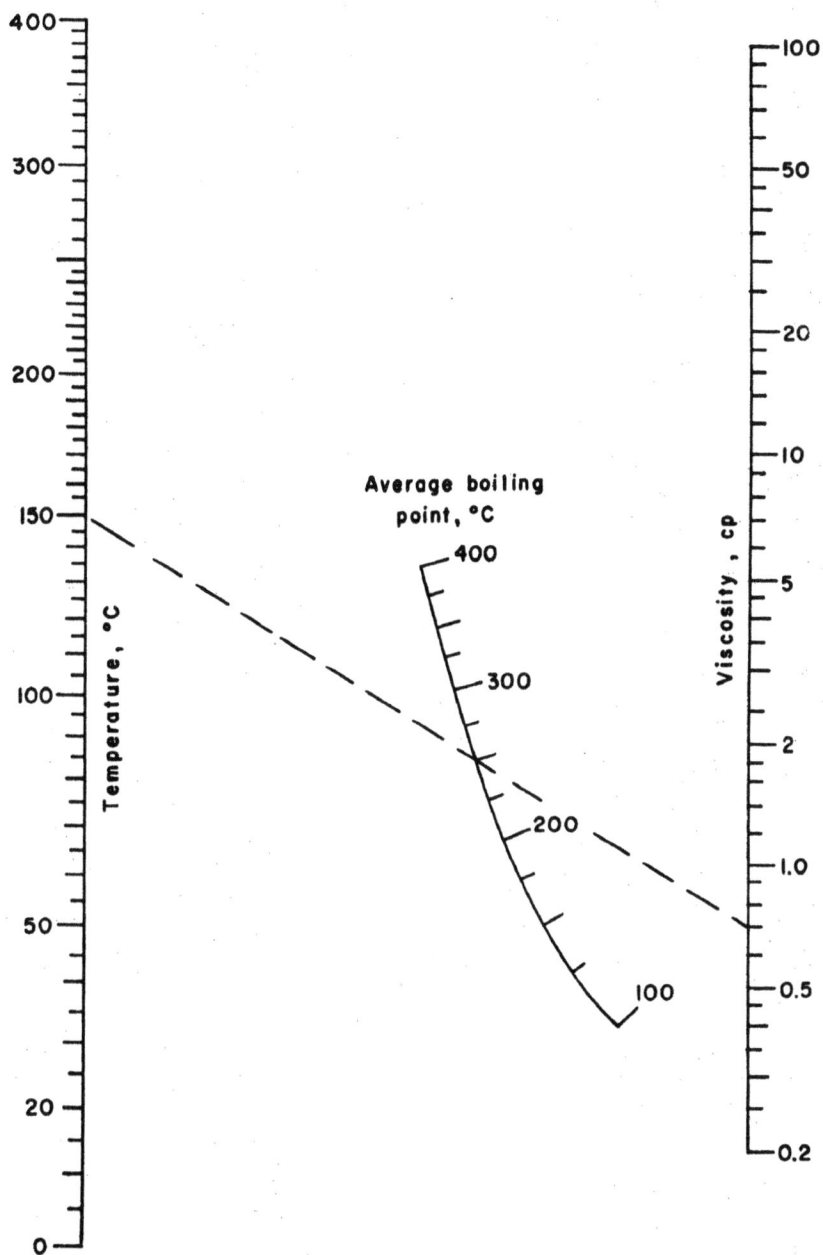

Figure 8-68

Author Index

Subject Index

A

Air, adiabatic expansion of, 166
  flow through rounded circular
    orifices, 26
  infiltration, 28
  lift, 332
Alkali equivalents, 314
Ammonia condenser, vapor composition in, 190
API gravities, asphaltic materials,
  156
Asphalts, characterization factors
  of, 202

B

Barometric condenser, water re-
  quirement of, 212
Benzols, aromatic content of, 192
Bleach liquor, flow rate of, 10
Blowdown, boilers and cooling
  towers, 196
Boiling point correction, 198, 200

C

Chemical addition, flow rates, 12
Coal, heating value of, 204
Concentration, 210
Conductance, thermal, of air
  spaces, 214
Cone, volume of, 216
Consistency of paper pulp, 95, 96
Conversion, of concentration units
  for petrol additives, 312

pressures, 316
Costs of pumping, 238
  scaling-up equipment, 236
Cryogenic refrigeration, 256
Crystallization, fractional, 220

D

Densities of alcohols, 139
  alkyl esters, 140
  asphaltic materials, 139
  fatty acids, 139, 140
  ketones, 139
  methanol-water, 148
  mineral acids, 139
  vapors at low pressures, 162
Diameters, equivalent, of tube
  bundles, 50
Dilution, 210, 240
Dished heads, volume of, 276

E

Efficiency, extractive, 310
  of heat exchangers, 308
Electrodeposits, thickness of, 222
Equilibrium, vapor-liquid, 296,
  298
Extruder, plastic, 320

F

Film coefficients for liquids, 64
Filters, high-temperature, gas bag,
  194
Flare stack contaminants, 224

www.ingramcontent.com/pod-product-compliance
Lightning Source LLC
Chambersburg PA
CBHW060808220326
41598CB00022B/2565